Investigations into the Trans Self and Moore's Paradox

Linda A. W. Brakel

Investigations into the Trans Self and Moore's Paradox

palgrave
macmillan

Linda A. W. Brakel
University of Michigan
Ann Arbor, MI, USA

ISBN 978-3-030-44644-4 ISBN 978-3-030-44645-1 (eBook)
https://doi.org/10.1007/978-3-030-44645-1

This Palgrave Macmillan imprint is published by the registered company Springer Nature Switzerland AG
The registered company address is: Gewerbestrasse 11, 6330 Cham, Switzerland

To Arthur Brakel, who continues to make it all worthwhile.

Acknowledgments

I credit and thank my husband, Arthur Brakel, for the initial spark catalyzing the inquiry and hence the book. Also, allowing for the very possibility of this book, I appreciate the work of an anonymous Palgrave Macmillan reviewer, and my editor, Lauriane Piette, both of whom were encouragingly interactive.

For invaluable theoretical and scientific information, I thank Justin Rhodes of the Beckman Institute at the University of Illinois. Additional acknowledgment for scholarly suggestions goes to Barb Smuts of the University of Michigan.

For steady listening, I am grateful to my friends, Barb Smuts and Michael (Spider) Abner, with a special note for powerful questioning from Michael. And in this category too I recognize my intrepid husband and life-love partner, Arthur Brakel.

Contents

Part I

Introduction

1
Introduction

Abstract The Trans phenomenon occasions vexing questions about the concept, *Self*. The book's catalyst is a Moore's Paradox question: Can a natally-assigned girl state without incoherence: "I am a girl, but I don't believe that"? Deeper questions arise: Ontological (what *Self* believes the natal-gender is "wrong?"), Epistemological (What does that *Self* know?), and Conceptual. This chapter provides the roadmap to avenues traversed toward addressing these matters. First, extant philosophical *Self* accounts are explored. Still seeking answers, the monograph turns to biology: An evolutionary proper-function account of gender/Trans-gender is developed. Then recent brain studies on Trans humans and sex-changing fish are reviewed. Finally, empirical Experimental Philosophy studies adapted from Classic Thought Experiments on *Self* are offered. Also introduced are "Contrast-Cases," where the *Self* questions are less problematic.

Keywords Trans phenomenon · Moore's Paradox · "Wrong gender" · Mainstream gender binary · Three fundamental questions—ontological, epistemological, conceptual · Contrast Cases

L. A. W. Brakel, *Investigations into the Trans Self and Moore's Paradox*,
https://doi.org/10.1007/978-3-030-44645-1_1

1 Prelude to the Project

This monograph will take up aspects of the Trans experience in an unusual way, investigating the philosophical concept of *Self*—specifically exploring how the phenomenon of Trans-ness challenges existing concepts of the *Self*. There is an important assumption embedded here: Namely, that it is change in the concept of *Self* itself, rather than discovering any irregularities in Trans persons' selves, that could advance a better fit between theory and reality. That said, let me introduce the project's several launch sites, indicative of the book's thoroughly interdisciplinary approach.

First, directly following Chapter 1, the current introductory chapter, Chapters 2–4 examine major categories of existing philosophical *Self* theories, demonstrating that they do not fare well in accommodating fundamental *Self* questions the Trans experience highlights. Chapter 2 explores four classical philosophical views of the *Self*—that of Descartes, Hume, Kant, and Freud. Chapter 3 takes up two classes of modern views—(1) the *Self* in self-constitution; (2) the *Self* or "me" I'm concerned with when self-survival is at stake. And Chapter 4 discusses selected gender-theorists on the *Self*.

Then in Chapter 5, the work takes a major turn, as an evolutionary "proper-function" account of gender and Trans-gender is developed and proposed, including a new ontological phenotype. This proper-function account links together some of the otherwise rather disparate views of the thinkers discussed in the prior chapters—and does so in a biologically relevant fashion.

Chapter 6, continues with biology, examining approximately twenty neuroscience brain research studies conducted over the last twenty-five years related to Trans- vs. Cis-gendered persons, and relevant to the proper-function analysis. I review the research, offer evaluative commentary thereupon, and conclude Chapter 6 with "proof-of-concept" analogies from sex-changing fish.

Chapter 7 consists of several empirical Experimental Philosophy (X-Phi) Thought Experiments, each adapted from four well known classical Thought Experiments on the *Self* introduced in Chapter 3. These Experimental Philosophy studies were designed for two purposes: (1) to provide some diagnostic data about individual Trans persons, potentially useful toward future interventions; and (2) to gather evidence for (or against) the notion of a new ontological phenotype posited in earlier chapters of the project. Chapter 8, Summary and Conclusions, offers not only a summary review, but also an outlook for the future.

* * *

In reading this Prelude, one will perhaps already have become aware that something most often present in works about the Trans experience, is absent here. Indeed, this project does not include the always interesting, often compelling, personal and clinical accounts of and by Trans persons. This is a purposeful omission as the current project is one that is more a nomothetic than an idiographic exploration. Many, perhaps most other authors do take up the Trans phenomena in a personal subjective and/or individual clinical fashion. Moreover, I am confident that many will continue to do so, offering the insights they gain in myriad diverse forms: clinical reports, biographies, autobiographical books, and essays. I admire these works, and moreover appreciate their aims, but realize that my potential contributions, if successful, must arrive from the different tacks outlined above, and elaborated on in the pages to follow.

Now before I introduce the motivation for the project, which revolves around several unanswered, but central philosophical questions forming the core of this monograph, there are two more contextual preliminaries to present. They follow immediately below under the heading of "General Considerations about Trans Phenomena".

1.1 General Considerations About Trans Phenomena

Other Cultures

Other cultures have frequently recognized more than two genders, i.e., persons who are considered outside of the ordinary mainstream binary of male or female, and instead represent some variation of a third gender. According to the Wikipedia entry for "third gender," there are a number of modern societies in addition to several historical ones with persons thought to have genders outside the typical binary genders, all acknowledged as people of one sort of so called "third gender" or another.[1] Also, even in our current western mainstream culture,[2] there are persons who are identified as fluid-gendered, mixed-gendered, unspecified gender, and non- or a-gendered. As important, interesting, and perhaps even predominant as these gender variants are, the focus of the present project will be centered on (1) denizens of the contemporary western mainstream, and in particular (2) those Trans persons who participate in the world of binary genders—male or female—but know and feel that their natal assigned gender is "wrong." Narrowing the scope even more, this project will deal with individuals without any genetic or gonadal anomalies, and without major psychiatric diagnoses. This admittedly restrictive framework is adopted with the understanding that the most extreme, stark, unambiguous cases can best provide the clearest understanding of the conceptual phenomena on the nature of the *Self* to be explored.

[1]The reference here is Wikipedia accessed on February 18, 2019 at https://en-wikipedia.org/wiki/Third gender. An incomplete list from this Wikipedia entry includes the following modern cultures, all of which recognize third genders of various types. There are three from Africa: Ethiopia, Kenya, Democratic Republic of Congo; five from the Asian-Pacific area: Indonesia, Polynesia, Micronesia, Japan, the Philippines; two in Europe: Balkans, Neapolitan Culture; at least two in present day Latin/South America: Mexico, Brazil; and many in American Indigenous Cultures, some under the umbrella term, "Two Spirit." There are historical cultures with third genders too. Here are just some examples: Several Indian Cultures; Mesopotamia; Thailand; Egypt; Old Israel; at least one ancient Mediterranean group; and a number of historic Mesoamerican cultures—the Maya, Aztec, and Inca. Of course, the Wikipedia entry just scratches the surface of the vital topic of non-binary genders. For those interested in a deeper look, *Transgender Warriors* (1996, by Leslie Feinberg, Boston: Beacon Press) is recommended as an initial reference source. (This suggestion owes to an anonymous Palgrave-Macmillan reviewer.)

[2]What I'm referring to as "our current western mainstream culture" is termed "western or North culture" by the gender-theorist Linda Alcoff (2006, Chapter 6, p. 8). Some of Alcoff's views will be discussed in Chapter 4, on the *Self* views of modern gender theorist philosophers. (I thank Professor Barbara Smuts (personal communication) for first bringing to my attention the need to reference the various sorts of differently gendered people, both in our current culture, historically, and worldwide.)

Fish

Many species of fish routinely change sex. According to marine biologist Richard Francis (1992, p. 1) "…sex change is particularly common in several species of coral-reef fishes…[In] marine teleosts generally, [sex changing] may prove the rule rather than the exception." In addition, several species have distinct morphs—natal males who become female-like in all but their gonads, and thereby are mismatched in terms of sexual gametes and gender as they develop. A model form of this de-coupling, as described by marine neuro-endocrinologist, John Godwin (2010, p. 209), includes so-called "parasitic male phenotypes" which while gonadally male, have the morphological and behavioral characteristics of females, not those of the "…large territorial male morph [with]…exaggerated morphological secondary sexual characters, [and] aggressive defense of a [breeding] territory…, and often conspicuous courtship behaviors." These parasitic morphs, on the other hand, are described by Godwin (2010, p. 209) as female-look-alikes, in that they are "smaller…and do not display the morphological [bodily] specializations [for courtship and territorial defense] exhibited by the larger males…(examples of which include coloration, fin extensions…)." Found in a number of diverse species—wrasses, parrot fish, sunfish, some salmonids, and plainfin midshipman (pp. 209–211)—Godwin demonstrates the operative power of evolutionary reproductive fitness success gained for alternative morphs via the example of sunfish "satellite males" (2010, p. 209): "Parental males are large…exhibit display coloration, and defend nest sites that females visit to deposit eggs…while satellite males mimic females in order to join spawning pairs and [in this different way] release sperm."

In species demonstrating complete sex change, as with those with alternative morphs, it is evolutionary reproductive fitness success that drives the system. It is my intention to introduce aspects of full sexual transition and gender/sex morph incongruity in fish in order to provide a template for analogous human phenomena. Specifically, I will propose a philosophical evolutionary proper-function analysis of gender and then the Trans-gender phenomena—an analysis, alluded to at various places in this work, and fully developed in Chapter 5. Beyond this, it is hoped that the proper-function analysis advanced, will not only provide a link to the many disparate elements of this project, but will prove, at the very least interesting, and perhaps even compelling insofar as sex-changing fish can serve a proof-of-concept role.

* * *

But, so as not to get out too far ahead, let me introduce the real stimulus for this work—the catalyst puzzle, and then the three motivating central questions—all of which I will elaborate upon just below.

2 The Questions—The Motivation for the Project; And Its Catalyst

This project seeks to investigate the philosophical concept of *Self*, given the idea that classical and modern versions alike seem challenged, frankly inadequate, in light of the experience of Trans persons. What follows then is first an account of the catalyst for this project, and then the fundamental philosophical questions motivating the exploration of the *Self* in light of the Trans phenomenon.

2.1 The Catalyst: A Specific Question—Moore's Paradox and Trans Phenomena

For the paradigm example, employed not only for this initial question, but in general throughout this work, let us take the case of a girl who feels and knows she is the "wrong" gender and will endeavor to transition to being a male person, a boy (or man). Hence, let us refer to this person as G $\rightarrow$ B (GB for easier reference). I will make the following important assumptions about GB's mental and physical health: GB is a child without genetic or gonadal disorders or ambiguities, and without any major psychiatric diagnosis—no psychoses, no major affective disorders, no attention deficit hyperactivity disorders (ADHD), and no pathological character or personality diagnoses. Another assumption concerns GB's socio-economic status: GB is situated in a family that is largely comfortable both emotionally and financially, and as a child has had only typical neurotic level problems, including some anxiety and depressive affect. Again, I take up this case before GB has undergone any physiologic interventions concerning gender.

With GB as the prototypic subject, here is how the Moore's Paradox question unfolds. Of great interest as one who proposes that certain Moore Paradoxical assertions need not be irrational or incoherent (Fileva and Brakel 2018)—e.g., a spider fearful person can affirm: "I know Huntsman spiders are benign, but I don't believe that"—I note that GB might rationally be able to state two forms of such Moore Paradoxical statements. In the positive form, GB could say: "I know I'm a girl, but I don't believe that." While, in the negative form, GB could say: "I know I'm not a boy, but I believe I am a boy." Further, GB could make these sorts of statements—statements that are best classified as neither delusional nor wishful—frequently and routinely.[3]

There is a potential variation, in that GB could aver: "I know I'm a boy; but I can't (quite) believe it." And although some suggest this version is more accurate, there are problems with the variation. Not only does the qualifier "quite" mitigate the strength of the belief, but more important, the second clause assertion, "I can't believe," is not actually a Moore-Paradoxical statement. "I can't believe"—i.e., not being able to believe, is itself subtly but significantly different from the starker attitude of not believing—"I don't believe."[4]

---

[3]While both the positive and negative versions are bona fide Moore-Paradoxical statements, it could be of potential interest, philosophically and/or psychologically, if empirical research found that one form of the assertion over the other was more likely endorsable. This notion will be taken up further in Chapter 7 on Experimental Philosophy Thought Experiments. (Further, it is hoped that with the publication of this monograph, researchers in clinics serving Trans-gender people might be motivated to run these no-risk experiments.)

[4]All manner of statements that seem readily able to dissolve the paradox of Moore-Paradoxical assertions turn out to be not genuinely Moore-Paradoxical. Take for instance, "I judge that it is raining, but I don't believe that." One could easily, and for a variety of reasons, not trust one's judgement. In this way this statement is not Moore-Paradoxical despite its superficial resemblance to the paradigmatic "It is raining, but I don't believe that." There is indeed a sizeable literature on Moore-Paradoxical statements, including works on the criteria for counting as such, but this matter is tangential to the current inquiry. (For those with a basic interest, the Fileva and Brakel (2018) article cited above, particularly the references section, might be of interest for a start.)

Returning to the original indisputably paradoxical Moore-Paradoxical statements described above, there is a looming and serious difficulty in accurately investigating Trans persons assenting—either to the positive or negative form—of these particular Moore-Paradoxical assertions. This problem is an intense political one: Persons wanting medical professionals to participate in transition treatments need to first demonstrate that they are "certified patients." This means they must fit the narrow diagnostic criteria as enumerated in the American Psychological Association's DSM-V (Diagnostic and Statistical Manual of Mental Disorders—Fifth Edition) for the diagnostic designation "Gender Dysphoria."[5] One of the criteria for being this sort of "patient" is a longstanding, in fact, life-long belief that the person in question has somehow been born with a wrong-gendered body, and as such is "exhibiting a strong and persistent cross-gender identification, which [includes] … the insistence that one is of the other sex."[6] Thus, on these instrumental grounds alone, for a person like GB who wants to transition to male-ness, the initial clause –"I am a girl" of the positive form of the Moore-Paradoxical assertive statement—"I am a girl, but I don't believe that"—would be highly problematic to

[5]Note that the diagnosis "Gender Dysphoria" or "GD" is a 2013 re-classification in the American Psychiatric Association's *Diagnostic and Statistical Manual of Mental Disorders-Fifth Edition (DSM-V)* of the older diagnosis, "Gender Identity Disorder" or "GID." GID was the mental disorder defined as the felt incongruence between the assigned biological gender and gender identity in the absence of any physical intersex condition. The current GD diagnosis includes this, but also adds and emphasizes the distress occasioned by the felt incongruence—an incongruence which owes to "being born in the wrong body"—is longstanding (perhaps as early as the advent of one's conscious awareness), and based on the conviction that the person in question is the wrong gender. Now, it is the case that the GD diagnosis itself may be in the process of substantial revision, as is evidenced by a recent (2018) ruling on a similar diagnostic category, "Gender Incongruence" by the World Health Organization (WHO). The WHO no longer considers "Gender Incongruence" a mental disorder. Instead, while recognizing that persons experiencing gender incongruence may (and often do) experience dysphoria, the condition is listed in the 2018 *International Statistical Classification of Distress* (ICD) as a "Sexual Health Condition." (This information is from an American Medical Association Daily Briefing, June 18, 2018.) Two recent New York Times op-ed columns also advance the notion of healthy normalcy for Trans persons. This can be appreciated even from the two titles. Barnard English Professor, Jennifer Finney Boylan's (January 9, 2019) adopts a teacherly tone, explaining to various skeptics that with regard to being Trans-gendered: "It's not a teenage fad. It's life." And Roy Richard Grinker, anthropologist at George Washington University, titles his December 7, 2018 op-ed with the unambiguous claim: "Being Trans is Not a Mental Condition."

[6]This partial sentence is a quote from the DSM Fifth Edition referenced by the philosopher Georgia Warnke (2009, p. 39).

endorse. GB would have to claim that the notion that "I am a girl" was never correct. Similarly, and for the same reason, the first clause of the Moore-Paradoxical negative assertive statement—"I am not a boy, but I believe I am a boy"—would be equally untenable for GB as a Trans persons who would want any medical interventions toward transition. GB would have to present as a person who long believed in GB's maleness, whatever evidence there was to the contrary notwithstanding.[7]

2.2 Beyond Moore's Paradox: The Three Foundational Questions About the Nature of *Self* Arising from Considerations of Trans Phenomena

In whatever fashion the Moore's Paradox question is decided upon (or if it cannot be settled), Trans experiences—e.g., from GB knowing that gender transition is desired; to GB becoming male; to GB being a Transmale—afford us a unique opportunity to explore the nature of *Self* in a deeper and more radical fashion. For example: What aspect (or aspects) of *Self* are involved in GB knowing (and feeling) that GB is (or at least is supposed to be) male and that therefore GB wants to be/become male? Additionally, what aspects of GB's *Self* knows this? Also, what is it exactly that GB knows, and with what evidence?

Now before I elaborate on these very questions—questions intended to delve deeply into the nature of *Self* itself—and while we are still in this preliminary introductory chapter, it is important to re-iterate that throughout this monograph the ensuing discussions on the nature of *Self*

[7]Demonstrating that there are legal as well as personal consequences when Trans persons do not fit the GD or GID diagnoses, note the following comments from gender theorist Cressida Heyes (2009), "…the medical apparatus of sex change (the precondition for legal reclassification) is available only to certain kinds of persons—those who suffer from the condition of GID [or GD]" (p. 146). Putting an even sharper point on this, philosopher C. Jacob Hale (2009, p. 47) states: "…transsexuals are forced to submit to an intricate, tightly woven set of regulatory regimes. If we desire medically regulated technologies, we must either insert ourselves into these regimes or forego adequate medical care and access to juridical mechanisms for changing sex and legal documents that require certification from physicians."

will be explored from a general nomothetic philosophical outlook. This, rather than from any individual subjective psychological viewpoint.[8]

With this in mind, directly below are the three fundamental philosophical questions about *Self* that arise with regard to persons like GB. As we quest for a more adequate philosophical concept of *Self*, these are vital foundational questions with which we surely must grapple:

1. The ontological question:

> Which aspect(s) of GB's *Self* knows that "her" female gender is "wrong?" Similarly, to which aspect of GB's *Self* does the original female gendered-self feel so wrong? This is an especially vexing question given that GB's objective natal bodily self appears to produce and perceive normal sensations in a regular empirical fashion. Then there are further questions along these same lines: Which *Self* aspect or aspects wants to, and then does, act to transition?

2. The epistemological question:

> Given all the evidence that someone deemed "female" receives, unremittingly, both from internal (bodily) sources and from one's diverse external environments—evidence accruing from birth onward regarding in particular "her" female gendered-ness—how and what does a person like GB know about the wrongness of this initial assignment? And what is it that constitutes GB's evidence?

3. The conceptual question:

> Can the epistemological and ontological dilemmas alluded to above shed light on the very concepts we have regarding the nature of *Self*? Is a change in the very concept of *Self* necessitated?

* * *

[8] I do this while acknowledging that the boundaries between the relevant psychological and philosophical disciplines are often not clear-cut, which is all to the good in my admittedly interdisciplinary view.

Now that the motivating questions have been set forth, and in light of the fact that this project is rather complex consisting of a number of varied but intersecting parts, a road map to the project is in order and follows directly.

3 Road Map: The Plan for the Project

3.1 Part I: Introduction

Chapter 1—Introduction

This chapter began with the section on "Prelude to the Project" which aimed to set out contextual matters. Next discussed was the particular but narrow catalyst—the Moore's Paradox question for Trans persons. Finally, the "Prelude" section closed with the true motivation for the project—the three foundational philosophical questions about *Self*, ontological, epistemological, and conceptual, that arise rather naturally when considering the Trans experience.

This introductory chapter continues now to provide the "Plan for the Project," a road map to guide the reader through the twists and turns this monograph will take. At this chapter's end, there will be three initial "Contrast Cases." Contrast Cases are so termed in that they are related to, but better understood than the Trans-*Self* case. Contrast Cases will be crosscut throughout the work as is relevant.

3.2 Part II: Extant Philosophical Views of *Self*

Several extant views of the *Self* are selected; all with the goal of evaluating how well each can fare toward answering the three foundational philosophical questions at the heart of this monograph—the ontological, epistemological, and conceptual questions about the nature of *Self*—problematized owing to the Trans phenomenon.

* * *

Chapter 2—Four Classical Philosophical Views of *Self*

In chronological order, the accounts of four seminal thinkers about the *Self*—Descartes, Hume, Kant, and Freud—are reviewed. The major points of each philosopher are outlined, then some flaws. Finally, I evaluate each of their concepts of *Self* in terms of the pressing foundational philosophical questions the Trans experience highlights.

Chapter 3—Modern Philosophical Accounts of *Self*

In this chapter a variety of modern accounts are considered, divided into two subsections: (1) The *Self* in self-constitution and (2) The *Self* at issue as the "Me" I care about when I care about my survival.

Under (1) The self-constitution heading, several views are explored, some familiar and some less so. First, and perhaps most familiar are (a) Pruning View of *Self*, the most famous of which is "the Endorsed *Self*." Other Pruning Views taken up are "the Rational *Self*" and "the True-*Self*." Next there are two accounts best characterized as (b) "Opponent-to-Pruner-*Self* views." Finally, there are (c) Conative Views of *Self*, which include three quite different *Self* accounts: (i) Integrated-*Self*; (ii) Central Cares Account of *Self*, and (iii) *Self* Constrained by Minimal Human Flourishing. As with the classical philosophical accounts, each of these modern views is examined with respect to the three foundational philosophical questions that arise owing to the Trans phenomenon.

Under subsection (2) The *Self* at issue when I care about my survival, I outline five different views, the first four of which constitute classical Thought Experiments: (a) Human Fission; (b) Body/Brain Exchange, i.e. Is my *Self*, my body or my brain/mind? (c) Tele-transporting, and the issue of Physical and Psychological spectra—"Relation-R" *Self* vs. Personal Identity *Self*; (d) Closest Continuer of *Self*; and finally, (e) My *Self* is my agency.[9] These accounts too, are explored with respect to the ontological, epistemological and conceptual fundamental questions

[9]Here is an advance road map note: The *Self* Constrained by Minimal Human Flourishing, and the views of the *Self* as agency, will be linked later in the project, specifically in Chapter 5, in which a proper-function evolutionary selective fitness account for gender and Trans-gender is offered. Also, it is helpful to keep in mind that many of the views outlined in the subsection

at the center of the project. But in addition, four of the Classical Thought Experiments are re-visited in Chapter 7, where they will be adapted to investigate Trans vs. Cis-*Self* issues in a series of empirical Experimental Philosophy (X-Phi) studies.

Chapter 4—Selected Gender Theorists Views

Surprisingly, the issue of *Self* (at least as I've framed it) does not come up as such within this huge and important literature. Nonetheless, I select four current philosophical views from Gender Theorists to discuss and evaluate, again with respect to the foundational philosophical questions about the *Self*. These are: (1) Gender as a Natural Kind; (2) Multiple-Meaning View of Gender and Concept Amelioration Project[10]; (3) Dispositional Account of Gender; and (4) Aspirational View of Gender Identity.

* * *

Each of the many accounts reviewed in Part II (Chapters 2–4), while seemingly able to explain aspects of the *Self* rather well, is not terribly useful in unraveling the mystery of *Self* manifest in the three fundamental questions raised by the Trans experience. Nor will combinations of the views resolve the puzzles. Thus, the next section (Part III) and the chapters (Chapters 5 and 6) within it, mark a radical shift in the monograph's direction. Realizing I cannot at this time answer the three fundamental questions of *Self* brought to light by the Trans phenomenon—not with classical, modern, or gender-theorists' accounts—I seek different sorts of questions—questions that can be answered, and with luck bear (at least) tangentially on the initial foundational questions at the core of this

---

on the nature of *Self* when self-survival is at issue, will be re-visited in Chapter 7, on empirical Experimental Philosophy (X-Phi) Thought Experiments.

[10]The Multiple-Meaning View of Gender and the Concept Amelioration Project will also fit with Chapter 5 proper-function evolutionary fitness success analysis. The proper-function account I offer there provides a key link for the project as a whole. In addition, the Multiple-Meaning View of Gender along with the Concept Amelioration Project allows (or at least hints at the possibility of) a new ontological category—a Trans phenotype. The existence of a Trans phenotype is a hypothesis I advance, and then explore more fully in Chapters 5–7.

project. And with this, the monograph turns toward biology—evolutionary biology and neuroscience brain biology.

3.3 Part III: Biological Considerations

Chapter 5—A Proper-Function Account of Gender: Bridge from Philosophical Views to Biological Matters (Including Fish)

Radically changing direction, this chapter integrates several of the modern philosophical views discussed in Chapters 3 and 4, along with other resources, in order to derive a proper-function account of gender and Trans-gender, with evolutionary selective reproductive fitness as the operative mechanism. Additionally, the notion of two new ontological phenotypes—Trans-men and Trans-women—will be introduced. Observational research both on certain fish species that change their sex and others that modify their morphological gender (alternative morphs) are presented, providing proof-of-concept type support to the proper-function analysis and the Trans phenotype idea.

Chapter 6—Brain Studies: Trans-persons and Fish

Continuing with biological concerns, several neuroscience brain studies (1995–2018) involving both Trans and Cis persons are outlined. Findings, as well as criticisms are offered. Brain causes vs. epigenetic brain effects of the Trans experience will be explored. This section also includes brain and behavioral research on sex-changing fish—particularly brain studies before, after, and during transitions—suggesting that this comparative species investigation can again offer substantial proof-of-concept evidence for the new ontological phenotype introduced in Chapter 5—Trans-man males; Trans-woman females—as well as suggest potentially useful data for future exploration.

3.4 Part IV: Future Investigations

Chapter 7—Experimental Philosophy (X-Phi) Studies

In Part IV, which consists solely of Chapter 7, several Experimental Philosophy (X-Phi) empirical experiments are presented. These have all been adapted from the classical philosophical Thought Experiments on the nature of *Self*, discussed earlier in Chapter 3. They include: Body/Brain Exchange, Tele-transporting with Percentage Change, and the Closest (Real) Continuer. The new studies offer the possibility of progress toward (a) providing empirical support for two new ontological phenotypic gender categories—Trans-men and Trans-women; and (b) improving diagnostic acumen in distinguishing between two groups of Trans persons: Those who feel that their body is the locus of incongruence and would be best transitioned to the other gender; and those who on the contrary believe that their brain/mind is the source of disconnect and in need of change.

Chapter 8—Summary and Conclusions

This chapter begins with a summary of the monograph's explorations and findings. It includes a brief recap of currently existing concepts of *Self*, both classic and modern, and a review of the ongoing problems in addressing the core foundational philosophical questions brought to light by the Trans phenomenon—the three fundamental questions, ontological, epistemological, and conceptual motivating this project. The proper-function account for gender and Trans-gender I derived is then revisited, as well as the idea of ontologically new phenotypes—Trans-men and Trans-women. This last notion not only returns us to the original Moore's Paradox dilemma, but also provides some insight into future developments and directions.

4 The Three Initial Contrast Cases

Before we get to Part II/Chapter 2, as promised, Contrast Cases will be introduced. Contrast Cases are cases that, while having significant issues similar to that of Trans persons, do not raise the same fundamental explanatory difficulties for existing concepts of *Self* that the Trans phenomenon does. There are three such initial cases to consider: *Contrast Case 1* concerns racial identity change; *Contrast Case 2* takes up persons suffering from anorexia and other body dysmorphias; and *Contrast Case 3* discusses non-binary gendered persons, including those who are fluid-gendered, mixed-gendered, unspecified gender, and non-gendered.

4.1 Contrast Case 1—Racial Identity Change

On the face of it, racial identity change cases seem quite similar to gender identity change cases. Whether it be black persons living (or attempting to live) as white, formerly termed "passing," or instances like the recent (2016) news-making Rachel Dolezal, a white woman, trying to live as black; these race change identity cases, while deeply embedded in highly problematic socio-political contexts, are not as philosophically vexing, at least not in regard to the matters explored here. Why not? Because the fundamental *Self* questions so alive in the Trans experience, do not occur here. This can be demonstrated in that Moore-Paradoxical assertions (of either the positive or negative sort) do not obtain for people attempting to change their racial identifications. People identified as white Caucasians through their growing-up years, especially when this is congruent with the racial identity of their biological parents, cannot without dissembling and deception sincerely assert, "I am white, but I don't believe I am white," even as the person in question choses to live as a black person. Nor can such a person state: "I know I'm not black, but I don't believe that I'm not black." These sorts of cases can be much better understood in terms of wish, desire, and denial—"I am white, but I wish I were black and I will live as though I were." And "I know I'm not black, and I believe I'm not black, but I will deny that knowledge, living as though I were black." Or, as perhaps was more prevalent in the early through the mid twentieth century America, "I know I'm not white, and I believe I am not, but I desire to be treated as white (as I know about white privilege and black disadvantage) and I will live as though I were white."[11]

4.2 Contrast Case 2—Anorexics and Other Persons with Body Dysmorphias

Unlike the situation with Racial Identity Change, people with anorexia (and other body dysmorphias) can indeed assert Moore-Paradoxical statements without incoherence, but they cannot do so without irrationality. An anorexic with a very low BMI (Body Mass Index) might honestly, but quite irrationally, claim: "I know I'm not fat, but I don't believe that" or "I know I'm too thin, but I don't believe that." (We took up a case like this in Fileva and Brakel 2018, pp. 9–10.) Similar, but more clearly caused by neuropathology and therefore secondarily irrational, are those patients suffering from one of the strange (and fortunately rare) family of ailments called, Xenomelia, Apotemnophia, Alien Body Part Syndrome, or Bodily Integrity Identity Disorder (BIID).[12] These patients believe that one or more of their own limbs or appendages feels alien, so much so, that in very extreme cases, the patient unceasingly strives to find a surgeon to amputate.[13] Such patients, despite the irrationality, can coherently and genuinely assert: "This leg is mine, but I don't believe it." However, coherent Moore-Paradoxical statements notwithstanding, and even while suffering from one of several patently irrational pathologies,

[11]More on the Rachel Dolezal situation, particularly as it reflects and contrasts with issues of Trans-gender and Transsexuality can be accessed by referring to "The Tuvel affair"—a controversy also involving an author and a feminist philosophical journal. One such cite is https://en.wikipedia.org/wiki/Hypatia_transracialism_controversy.

[12]For more on these syndromes, see Upadhyaya M. A. and Nasrallah H. A. (2017). There, the authors report that the neuropathology most demonstrated and documented involves changes to the thickness of the right parietal cortex. Consequent to this, the clinical symptoms most often affect the left lower leg.

[13]Interestingly, one could consider autoimmune diseases a cellular analog. There, certain self-tissues are, due to various causes, mistakenly tagged as non-self, with attacks by the immune system itself to follow.

the fundamental philosophical *Self* questions that do arise for Trans persons are not at issue here. Note, for instance, that although in anorexics there is a continuum of change to the body—and it can be massive—throughout the process, the person in question is almost never moved to change his/her name. It does seem then for anorexic persons, and even for patients with alien appendage BID syndrome (or other body dysmorphias), while there are significant, even extreme bodily changes, the problematized foundational *Self* questions highlighted by the Trans phenomenon do not arise.

4.3 Contrast Case 3—Non-binary Gendered Persons—Those Who Are Fluid-Gendered, Multiple-Gendered, of Non-specified Gender, and No Gender

There is one more initial Contrast Case that should be referenced as distinct from the sort of Trans person considered in this work. Actually, there are four different types within this category. These include persons who (1) identify as non-or a-gendered; (2) those who are fluid-gendered, including those who sometimes identify as male and sometimes female; (3) those who consider themselves multi-gendered in a mosaic, mixed or blended way; and (4) those who feel themselves to be of an unspecified gender. All four types might (or might not) be categorized as "third gender," and the referring pronouns could be "they" in fluid-gendered and multi-gendered persons, and "xi" in any of these four types. These all comprise Contrast Case 3 because Moore-Paradoxical statements do not obtain: So, gender fluid Person-A could say: "I am a natal boy, but that's not all that I am, I also identify as a girl sometimes." And multi-gendered Person-B could assert: "I am a natal boy, but although I am partly boy, I am also partly girl." And in a similar fashion, again without any Moore-Paradoxical type statement, non-gendered Person-C could assert: "I am Person-C, a person without a gender." Finally, non-specified Person-D might say: "I was born a boy, but though I am not a girl, I am a third-gender."

Additionally, in all of these cases there is another important difference from binary embracing Trans persons. The non-gendered person, for instance, could be seen as roughly analogous to the Alien-Body Part patient, absent the neuropsychiatric brain pathology. In both cases, an aspect of oneself—gender, in the non-gendered person, the body part in the Alien-Body case—is renounced or denied, but the puzzling foundational questions occasioned about the *Self* in the binary Trans case do not arise. This is likewise true for those in the other three types of non-binary gender categories—fluid-gendered, multi-or mixed-gendered, and unspecified gender: The philosophical questions about the *Self* do not occur, because for all of these non-binary persons there is no "wrongness of *Self* (or aspect thereof)" with which to contend.

* * *

Now, that the project's plan has been outlined, and the initial Contrast Cases presented, let us precede, without any more delay, to Part II—Extant Philosophical Views of *Self.* This begins with Chapter 2, four classical philosophical accounts.

Bibliography

Alcoff, L. (2006). *Visible identities: Race, gender, and the self.* New York: Oxford University Press.

Fileva, I., & Brakel, L. A. W. (2018). Just another article on Moore's Paradox: But we don't believe that. *Synthese,* 1–15. https://doi.org/10.1007/s11229-018-1703-1.

Francis, R. (1992). Sexual lability in teleosts: Developmental factors. *The Quarterly Review of Biology, 6,* 1–18.

Godwin, J. (2010). Neuroendocrinology of sexual plasticity in teleost fishes. *Frontiers in Neuroendocrinology, 31,* 203–216.

Hale, C. (2009). Tracing a ghostly memory in my throat: Reflections on FTM feminist voice and agency. In L. Shrage (Ed.), *You've changed: Sex reassignment and personal identity* (Chapter 3, pp. 43–65). Oxford: Oxford University Press.

Heyes, C. (2009). Changing race: Changing sex: The ethics of self-transformation. In L. Shrage (Ed.), *You've changed: Sex reassignment and personal identity* (Chapter 8, pp. 135–154). Oxford: Oxford University Press.

Upadhyaya, M. A., & Nasrallah, H. A. (2017). The intense desire for healthy limb amputation: A dis-proprioceptive neuropsychiatric disorder. *Annals of Clinical Psychiatry, 29,* 125–132.

Warnke, G. (2009). Transsexuality and contextual identities. In L. Shrage (Ed.), *You've changed: Sex reassignment and personal identity* (Chapter 2, pp. 28–42). Oxford: Oxford University Press.

Part II

Extant Philosophical Views of *Self*

2

Four Classical Philosophical Views of *Self*

Abstract With the aim of addressing the three fundamental philosophical questions—ontological, epistemological, and conceptual—arising owing to the Trans phenomenon, the *Self* views of Descartes, Hume, Kant, and Freud are reviewed. Each account is evaluated both for strengths and problems, with implications for the philosophical *Self* questions of central importance brought to light by the Trans phenomenon.

Keywords Classic *Self* views · Descartes · Hume · Kant · Freud

1 Descartes

1.1 General View on *Self*

"Je pense donc je suis," or "I think therefore I am," or "Cogito ergo sum." For Descartes this succinct expression, in any language, is the core of a quite brilliant proof of existence for whoever asserts it. Moreover, Descartes extends this assertion, fending off radical skepticism as follows:

L. A. W. Brakel, *Investigations into the Trans Self and Moore's Paradox*,
https://doi.org/10.1007/978-3-030-44645-1_2

I think therefore I exist.
I doubt therefore I exist.
I doubt that I exist,
therefore I doubt,
therefore I exist.

Descartes also has an answer to any evil deceiver:

If I am deceived that I exist,
therefore I am deceived,
therefore I exist.

Here is how Descartes puts it (Mediation II [1641/1969, p. 150] in *Meditations on First Philosophy*): "But I was persuaded that there was nothing in all the world, that there was no heaven, no earth, that there were no minds, nor any bodies: was I not then likewise persuaded that I did not exist? Not at all; of a surety I myself did exist since I persuaded myself of something [or merely because I thought of something.] But there is some deceiver or other, very powerful and very cunning, who ever employs his ingenuity in deceiving me. Then without a doubt I exist also if he *deceives* me [Descartes' italics]…"

For Descartes, any activity of awareness or consciousness, including whenever one "…doubts, understands, [conceives], affirms, denies, wills, refuses, which also imagines and feels (p. 153),"[1] secures existence for the being that is aware or conscious. Note that the awareness at issue need not be restricted to self-awareness (although it is possible that Descartes would not go along with me here). To make my case, I contend that in so far as the dog is seeing and chasing the squirrel, the dog is aware of the squirrel. And as being aware of the squirrel is being aware, the dog is aware. Therefore, the dog exists. Similarly, I need not be self-reflectively conscious of myself, to be conscious of various stimuli, thereby fixing the proof of my existence.

One truly wonderful aspect of Descartes' accomplishment is that he has taken an epistemological claim—he knows he exists—and with it establishes the ontological reality—the is-ness of his existence. This is as

[1]These are Descartes' brackets and italics, at least as is suggested by translator.

remarkable as it is unassailable, but this is not the whole story as some of his next assumptions are more questionable.

1.2 Problems with Descartes' View

In particular, whereas the original epistemological argument leads to an ontological truth—the existence of "I," Descartes' next move demonstrates a conflation of his epistemological given with an ontological one, as he contemplates the ontological essence of I/me. He starts benignly—beyond thinking, there is a thinking substance, I. Not transcendental, the I is an ego, consisting of "…the I which is left when everything other that the 'thinking' has been thought away" (Coplston 1963, Vol. 4, pp. 106–107). But trouble arises, strikingly with the very same content used for the cogito proof. Since Descartes can doubt that he has a body, he can imagine himself with no body; but because he cannot doubt the existence of his mind, he cannot imagine himself without a mind. Thus, given that the essence of any thing, X, is that which is necessary and indispensable for X to be X, Descartes has "shown"[2] that the nature/essence of I/me is as a thinking substance (thinking subject may be the better term), and further, a thinking substance that is immanent and non-substantial, and separate from the body. This argument is more fully explicated in Malcolm (1965, pp. 315–338).

The problem here is that Descartes is wrongly extending a valid epistemological argument about ontological existence to an invalid epistemological argument about ontological essence. Yes, my ontological existence can be claimed from my epistemological certainty—I exist, even as I imagine or think or doubt that I do not, because in the very imagining that I do not exist, I do exist as one who doubts or imagines or thinks. But, I and Descartes, can both be quite wrong in our epistemological thoughts, imaginings, and knowings about what constitutes the ontological essence of ourselves; and moreover, we can be quite wrong

[2]Although Descartes presents a version of this argument in Meditation II (1641/1969, pp. 149–157), he does not claim to have "proved" it until Meditation VI (1641/1969, pp. 185–199), in which God is the ultimate source of the clear and distinct presentations necessary for Descartes' proof.

about the nature of ourselves as thinking things. Whereas Descartes' thinking substance is non-material and separate from the body, I have a token reductive physicalist account of the mind/body.[3] As such, I hold that every thinking event is a body-event, as it is a mind-mental/body-brain event, and I cannot imagine it elsewise. That I cannot imagine it any differently, is no guarantee that I'm right about the embodied mind, just as Descartes' imagination can lead him to no valid claim the non-substantial nature of the mental. Indeed, the very essence of human thinking could be non-conscious cognition, rather than the conscious imagining Descartes posits as essential. In short, our ontological essences cannot be determined by our own imaginations, nor are they likely even determinable by our epistemological "knowledge" of ourselves. Descartes' epistemological argument for ontological existence argument, extending to ontological essence, is unwarranted.

1.3 Implications of Descartes' View of *Self* for the Trans Derived Questions

Starkly and simply put, Descartes' view of the *Self* provides no insight into the nature of the problematized *Self* concept in light of the fundamental questions brought to the fore by the Trans experience. However, perhaps surprisingly (and tangential to the vexing questions at issue), a Trans person might find a certain type of limited comfort in Cartesian dualism. Thus, GB, a natally-assigned girl who wants to and will transition to being male, could well express the following: "I doubt that I am a girl; but in doubting that, as would be the case in doubting anything, including my very existence, I've demonstrated that I exist." Indeed, GB's *Self*, her/his "I" or "me" can find unproblematic subjective first-person truth in the family of assertions characterized by I think/I doubt. GB might in this way assert: "I think, doubt, and even negate many things, especially about myself/my body, but therefore in this thinking, doubting, and no less in this negating, **I** am, I **know** that I am." At every stage of her/his transition and her/his life, the existence of GB's thinking

[3]My view, Diachronic Conjunctive Token Physicalism (DiCoToP) is presented in Brakel (2013, Chapter 3, pp. 41–92), "The Ontology of Psychology."

substance *Self* can be asserted with certainty, separate from the body's status, extreme bodily alterations notwithstanding. Good, as far as this goes; but it only goes so far. Thus, neither the Cartesian cogito argument, nor Descartes' further view of the essential ontological *Self*, has provided any progress toward answering the puzzling *Self* questions underlying the Trans phenomenon, including what might be required to revise our concept of the *Self*.

2 Hume

2.1 General View on *Self*

Hume advocates for the *Self* as a bundle of percepts, with the bundle composed of a collection of individual percepts. Here is his famous quotation (1739/1978, p. 252): "For my part, when I enter most intimately into what I call myself, I always stumble on some particular perception or other, of heat or cold, light or shade, love or hatred, pain or pleasure. I never can catch *myself* at any time without a perception, and never can observe any thing but the perception." Hume continues (p. 252):

> If any one, upon serious and unprejudic'd reflection thinks he has a different notion of *himself*, I must confess I can reason no longer with him. All I can allow him is, that he may be in the right as well as I, and that we are essentially different in this particular. He may, perhaps, perceive something simple and continu'd, which he calls *himself*; tho' I am certain there is no such principle in me…I may venture to affirm of the rest of mankind that they are nothing but a bundle or collection of different perceptions, which succeed each other with an inconceivable rapidity, and are in a perceptual state of flux." [Hume's italics]

For Hume, what he deems the "illusion" of a single simple *Self* is afforded by the human imagination's tendency to link impressions via the principles of association. These principles, including resemblance, contiguity, and causality, on the Humean account do not inhere in the reality of what is perceived. Nor, says Hume should "…we suppose the whole train of perception to be united by an identity…" Rather, the identity of a person, a *Self*, is not something that "…really binds our perceptions together…[but] only associates their idea in the imagination" (p. 259).

Hume is convinced that the imagination, which binds together the perceptions by the associative principles of resemblance, contiguity, and causality (this last, itself an imaginative linking of what is merely observed to be a "constant conjunction" of impressions), does not do its binding work as a function of an active unified mind. Indeed, according to philosopher Robert P. Wolff (1960/1966, p. 104), for Hume, these associative principles of the mind, result not from true mental activity but instead "a gentle force" from "the nature of the ideas themselves."[4] Continuing with his explication of Hume's view Wolff resumes (p. 104): "The imagination, if subjected repeatedly to the gentle force of association, develops certain *habits or customs* [italics are from Wolff and attributed to Hume]. It comes to anticipate the conjunction of perceptions which past experience has exhibited…as Pavlov's dog would begin to salivate at the sound of a bell…"[5]

Beyond this, and bearing more directly on the question of *Self*, Hume contends that the bundle of percepts, taken together, cannot constitute a singular mind, a singular identical *Self*, or in Hume's terms, a "personal identity." He says (1739/1978, p. 259): "The identity which we ascribe to the mind of man is only a fictious one…we suppose the whole train of perceptions to be united by an identity, [but] a question naturally arises concerning this relation of identity; whether it be something that really binds our several perceptions together, or only associates their ideas in the imagination…what has been already prov'd at large, that the understanding never observes any real connexion…" As is true for objects, Hume maintains (p. 262) for "personal identity…the relation of parts gives rise to some fiction or imaginary principle of union…"

---

[4]Although the "gentle force" originates with the "nature of the ideas themselves," it is also the case that for Hume, resemblance, contiguity, and causality are not inhering in the reality of what is perceived. Some interaction between the ideas themselves and our human proclivities, albeit not active mind proclivities, must be at work for the Humean position.

[5]In Brakel (2013, Chapter 2, pp. 13–37), "Extinction Phenomena: A Biologic Perspective on How and Why Psychoanalysis Works," contra Hume, I suggest that associations are the result of true mental activity, and that the associative principle underlies conditioning, including classical conditioning.

The notion of one's mind or *Self* as fictious is an eliminativist view of the *Self*. The basis of Hume's eliminativist view of the *Self* is discussed by Penelhum (1955/1966, pp. 216–217) as follows. Penelhum claims that Hume believes that almost everyone is subject to a "...confusion between two ideas: (a) that of an object which persists throughout a length of time without change or interruption—this is the idea of identity; (b) that of a succession of related objects—this...a case of diversity." But Penhelhum (pp. 227–228) goes on to hold that the confusion is really Hume's confusion and can be cleared up easily, by realizing that "...the pairs of expressions, (a) 'numerically the same' and 'containing many parts' and (b) 'numerically the same' and 'changed,' are not pairs of contradictories."

Hume's view of *Self* and a singular unified mind has also been characterized as a reductionist view. Penelhum (1955/1960, p. 219), for instance, holds that for Hume human beings are "...nothing but the series of *their own* [Penelhum's italics] sensations, feels, dreams, images, and the rest." Nelson Pike's take (1967, pp. 164–165) renders Hume's reductionism even more clearly: "Hume seems to have thought that statements about the *activities* [Pike's italics] of mind [self] (such as observing, remembering, believing, etc.), whether these activities are activities with respect to physical objects or mental states...are ultimately reducible to statements affirming the presence of certain specialized perceptions in the collection of perceptions making up the mind [self]." In other words, according to Pike (p. 165) "...for Hume, the mind [self] does not *do* anything—it *includes* [Pike's italics] things."

2.2 Problems with Hume's View

There are serious problems with the Humean account of *Self*, whether it is considered an eliminative or a reductionist view. Taking up the reductionist account first, let's examine again mental activity. Given that Hume holds that impressions/perceptions are linked according to resemblance, contiguity, and causality, an active associative mind seems necessarily implied in order to constitute, even to recognize among the enormous cascade of incoming percepts, some of them as similar, adjacent, and/or constantly conjoined to one another. Neither the "gentle force" from the "nature of the ideas themselves," nor the passive inclusion of specialized perceptions in a collection constituting the *Self*, seem at all adequate to the task.

Second, and of no less importance, here are a few pointed and difficult-to-handle questions for Hume, the eliminativist, asserting, as he does, that the *Self* and personal identity are fictions: Given that we humans continue to embrace a unified personal identity and a singular *Self* (both mistakes in Hume's view), we must ask Hume: *Who* if not one's *Self*, holds to this fictional idea of self? In other words, to *whom* does a unified, singular personal identity occur to as an idea? Certainly, the idea of a unified *Self* could not be attributed to a bundle or sequence of impressions. Moreover, Hume should be asked: What aspect of *Self* is involved when you are looking inside yourself and always stumble into a particular perception or impression. Rodrrick Chisholm (1969/1994, pp. 97–98) addresses this point: "If Hume finds what he says he finds, that is to say, if he finds not only perceptions, but also that *he* finds them…[then] there is *someone* who finds them…[Chisolm's italics]."

Joining both criticisms above, there is a rather snappy and totally apt comment that I believe originated with Anscombe, although I cannot find the reference. Nonetheless, I will paraphrase or quote it now, with appreciation to whomever it owes: Hume's *Self* is that which does the actual and active stumbling when he stumbles into particular percepts whenever he goes looking for his *Self*.

Interestingly, one could view Hume's problems as similar to Descartes' in the sense that both philosophers extend epistemological certainty to ontological claims. Since Descartes is epistemologically certain only of his thinking self, but not of his bodily self, he claims he is essentially ontologically a thinking being only. Hume, in looking for his *Self*, finds only particular impressions—impressions that then are collected and bundled. He is on firm ground, epistemologically, knowing these bound impressions. But then Hume takes what seems his unwarranted ontological step. Without addressing concerns regarding (a) which *Self* (or aspects thereof) is doing the looking inside himself; (b) which *Self* (or self-part) is stumbling upon particular impressions, and then finally; (c) which *Self* (or self-aspect) is collecting and bundling these perceptions,

Hume draws the conclusion that the *Self* as a unified singular entity is a fiction. And this ontological conclusion, derived as it is from epistemological facts, seems not only invalid, but more colloquially, wrong-headed.

2.3 Implications of Hume's View of *Self* for the Trans Occasioned Questions

The Humean account of *Self*—the *Self* as a fictional entity constructed by bundling separate perceptions in accord with principles of associations—does not help address any of the questions of *Self* raised in light of the Trans experience, questions that do not arise when considering Cis persons. Indeed, Trans persons looking inside for the *Self* bump into singular impressions just as Cis persons do. Hume's account can do no work here.

But, there is something more, and potentially of great interest: One can appreciate a striking and surprising similarity: Just as we would like to ask Hume—"To what *Self*, or aspect(s) thereof, do the bundled impressions/perceptions report, such that what you call the 'fictional self' can emerge?"—we would like to understand the following for Trans persons: "What *Self*, or aspect(s) thereof, realizes that the impressions/perceptions received, are received by a *Self* that is differently gendered from the natally assigned "wrong" one?" And now a final Hume question, a speculative one directed toward me and the readers: Can this parallel between a puzzle in Hume's notion of *Self* and the ontological question for Trans persons, prove useful?

3 Kant

3.1 General View on *Self*

Kant's understanding of the *Self* is quite complex, but his objections to the views of both Descartes and Hume provide a place to start. Kant (1781A/1965) in the *Critique of Pure Reason*, holds that Descartes' claim for the *Self* as a non-material thinking substance is untenable. "It is obvious that in attaching 'I' to our thoughts we designate the subject…, without noting in it any quality whatsoever—in fact without knowing anything of it either by direct acquaintance or otherwise" (A 355, p. 337). Likewise, the obverse claim to the effect that the *Self* should be characterized as an embodied empirically knowable entity would be equally problematic.[6] Thus, Kantian scholar philosopher Beatrice Longuenesse (2017, p. 102) succinctly states Kant's objections both to (a) Descartes, with his conclusions about the ontological nature of the *Self* as immaterial, and to (b) the embodied *Self* anti-Cartesians: For Kant, "It [the 'I think'] tells us neither that the referent of 'I' is *only* [Longuenesse's italics] a thinking thing, nor that it is anything besides a thinking thing."

According to Kant, Hume's account of *Self* as a bundle of perceptions is equally flawed. Kant in fact turns Hume's views upside down. Recall that for Hume, the bundling of a subject's singular individual experiences, is the best anyone can do toward constituting a *Self*—albeit a fictional one, comprising only a collection of separate experiences. Kant's view is radically opposite. He maintains that subjects could not even have any experience at all, without what he terms the "Transcendental Unity of Apperception." Remarkable analog to current neurophysiologic views of "the binding problem"[7]—i.e., the necessary combining of many simultaneous registrations from a myriad of perceptual modalities in order to even constitute objects as objects, including on a much more complex level, ourselves as objects and selves—the Transcendental Unity of Apperception (TUA) is a (perhaps the) central feature of Kant's epistemology as well as his understanding of the *Self.*

Longuenesse (2017, p. 89) explains the Kantian relations of the TUA, first to the "I", then to the "I think", and then to the "I" again: "What puts us in a position to use 'I' is *not* any particular experience but our capacity to bind and conceptualize representations, a capacity that makes *experience possible* rather than being derived from experience." Exactly the reverse of Hume's notion, Descartes' view is contested again too, in

[6]Note that at least a part of Kant's phrase "without knowing anything of it…by direct acquaintance…" likely refers here to the absence of empirical evidence for the *Self*—evidence which for him is itself not possible.

[7]See for instance, Edelman (2003) for more on the neurophysiology of the binding problem.

a few steps. So, according to Longuenesse (p. 104), "…Kant goes on to describe 'I think' as the *expression* of the unity of apperception: 'I can grasp [all my representations] together, as synthetically combined in an apperception, through the general expression **I think**' (B 138)." She continues (p. 107):

> …'I think' …[is] an *expression* of the transcendental unity of apperception [TUA] ….[But] 'I' can be equated neither with the transcendental unity of apperception nor even with 'I think.' Rather, it is an inseparable component of the proposition 'I think.' As such, many passages [e.g., B130, B153-4] indicate that Kant takes it [the'I'] to *represent* an entity, whatever entity that might be. [Bold emphasis above is Kant's; those in italics owe to Longuenesse's; those underlined are Brakel's]

To this point, and beyond his criticisms of Descartes and Hume, what account of *Self* can we attribute to Kant? Since, and in so far as, we do have experience(s), for Kant, we have an argument—a transcendental argument—for the existence of the Transcendental Unity of Apperception. And, if there is a TUA, there is an existing subject, a *Self*, one we can know, whose multiple representations the TUA functions to bind. Thus, Kant has secured ontologically and epistemologically a subjective *Self*, one with bound representations constituting experiences. This is an important achievement; but, so far this is a *Self* that is only synchronic, in other words a *Self* that is an extant subject only at-a-particular-time. A diachronic, numerically identical *Self*, one that is both existing over time, and as an objective entity, remains to be argued for. Here is how Kant (1787B/1965, p. 369) puts both what's been accomplished and what is still unsettled: "That the 'I,' the 'I' that thinks, can be regarded always as *subject*, …must be granted…but it does not mean that I, as *object*, am for myself a…*substance*" (B 407). Then elaborating Kant continues (B 408):

> That the 'I' of apperception, consequently in every thought, is a *singular*…and so refers to a logically simple subject, lies already in the concept of thinking, and is consequently an analytic proposition; but that does not signify that the thinking I is a simple substance, which would be a

> synthetic proposition. The concept of substance is always related to intuitions, which in me cannot be other than sensible, and hence must lie wholly outside the field of understanding.... [Kant's emphases throughout this paragraph]

It is clear then that Kant's next task is to provide proof for (or at least argue for) a numerically identical *Self*—a self that exists diachronically, over time, and is both subject and object. Kant (1787B/1965, pp. 244–247) does exactly this in the "Refutation of Idealism" section of The Critique of Pure Reason, specifically B 275–B 279. In most general terms, Paul Guyer (1992), Kant scholar, states that "Kant clearly perceived that there was some inescapable connection between self-knowledge and knowledge of objects, and ...both self-knowledge and knowledge of objects ... necessarily involved logical structures as well as empirical inputs" (p. 155). So, to begin Kant recognizes that each user of "I" considers all of the impressions/perceptions and intuitions experienced as belonging to a diachronic numerically identical subjective self—they are "mine." More specifically we are aware of these perceptions, etc. occurring sometimes simultaneously, but more often in various sorts of sequences. In other words, "in time." But, says Kant in B 275–276 (1787B/1965, p. 245):

> All determination of time presupposes something *permanent* in perception. This permanent [something] cannot, however, be something in me, since it is only through this permanent [something] that my existence in time can itself be determined. Thus, perception of this permanent [something] is possible only through a *thing* outside me...and consequently the determination of my existence in time is possible only through the existence of actual things which I perceive outside me...In other words, the consciousness of my existence is at the same time an immediate consciousness of the existence of other things outside me. [Kant's emphases throughout this paragraph]

Now, it is a short step to from our perceiving "actual" objects existing over time → to *Self* as a unified subject over time → to *Self* as a numerically identical "actual" object. Longuenesse (2017, p. 184) states the case well: "Kant is quite explicit about the fact that our representation of objects in the physical world depends not only on unifying representations under the unity of apperception…but also on our consciousness of our own position, as a physical thing, among other physical things." She continues (p. 184), that for Kant "…our experience of the spatial position of material things in space and our experience of our own spatial position with respect to them are mutually conditioning." Kant in (a slightly earlier subsection), the "Third Analogy" (B 261) refers to both objects in the physical world, and our bodily selves positioned in space perceiving these objects as he maintains (1787B/1965, p. 236): "…the relation of substances in which the one contains the determinations the ground of which is contained in the other is the relation of influence; and when each substance reciprocally contains the ground of the determination of the other, the relation is that of …reciprocity." Longueness (2017, p. 184) remarking on an adjacent passage of the "Third Analogy" clarifies: "Kant means to indicate that we are conscious of ourselves not just as thinkers…but also as empirical entities: living, sensing, thinking entities located in space and time."

The mutual dependence in knowing the existence of external objects and that of an ongoing numerically singular *Self*, one that is both a subject and an embodied object, is a Kantian turn that one cannot help but admire in its sweeping scope. Likewise awe inspiring, is Kant's emphasis on the experience granting power of the TUA. As Longueness (2017, p. 79) remarks, according to Kant, the TUA, often operating outside of consciousness, is "…at work throughout all our activities of apprehension, reproduction [memory], and recognition, as the necessary conditions for those activities to yield mutually consistent concepts applicable to one world of empirically given objects…[Indeed it] is the overall condition of any objective representation of the world…" From here, and as discussed above, considering the "I think" as a subject's conscious expression of the TUA—the TUA which often works implicitly—seems quite tenable. But, do problems arise when Kant extends this view? Does he go too far?

3.2 Problems with Kant's View

Kant claims "It must be possible for the 'I think' to accompany all of my representations; for otherwise something would be represented **in me** which could not be thought at all, and that is equivalent to saying that the representation would be impossible, or at least nothing **to me** [Brakel's emphases]….As *my* representations (even if I am not conscious of them as such) they must conform to the conditions under which they *can* stand together in one universal self-consciousness, because otherwise they would not all without exception belong to me. [Kant's italics]" (1787B/1965, §16 B Transcendental Deduction, B 131–132, pp. 152–153). A potent passage with many important implications; the problem it introduces is the necessity of the "I think" (moreover, the **conscious** "I think") as the sole **expression** of the TUA—the very binding force that secures any objective representation of the world. [Brakel's emphases] But obviously, many non-human ani mals have objective representations of the world, without the possibility of "I think" appended—neither the conscious "I think" nor a non-conscious version of "I think." Thus, to deal with this problem, we arrive at Contrast Case 4—Non-human Animals.

Contrast Case 4—Non-human Animals

Let's take most mammals and birds as the default reference class of non-human-animals. There is no dispute from Kant that these non-human animals have both "inner sense" and "outer sense"—these being Kant's terms for bodily registrations on the one hand, and impressions/perceptions from their outside environments, on the other. But Kant claims that for both inner and outer senses non-human animals have representations that are "in them" rather than representations that are "for them," as they are for humans. The difference, according to Kant is that animals lack not only the possibility of appending "I think" to every representation, but also the judgement that this expression signifies; namely, one that is required to put impressions etc. under concepts (intuitions). On this further matter, Longuenesse (2017, p. 198) characterizes Kant as holding that there are "…different types of…discriminating capacity; one that is based on the mere quality of the sensations; the other that is based on a higher order act that calls on the unity of apperception [TUA]...But the mere qualitative vivacity does not suffice for consciousness of the differences...such as ...generation of concepts."[8]

I don't agree with Kant here.[9] Many non-human animals may in fact lack reflective consciousness (i.e., being aware of being aware of something), as well as other "higher-level" cognitive capacities (those termed "secondary processes" by Freud and "system two operations" by modern dual process cognitive psychologists), which include, for example, the capacity to recognize the subjectivity and intentional view point of other beings. But it is equally evident that most non-human-animals function with concepts, and could not survive without them. Animals navigate the world with the ability to distinguish themselves as material objects from other material objects. Examples are everywhere and include the successful predating of predator animals, predation avoidance for prey animals, mating for every species, and the even more basic skill of not crashing into things. And even this sort of material differentiation of self from other objects is an important conceptual capacity. Now one could argue that all of the so-called concepts utilized by non-human-animals are linked to sensations produced in them. Kant makes this claim exactly. Longueness (2017, p. 198 n.25) quotes Kant to this effect (from a 1762 essay[10]): "The dog differentiates the roast differently from the bread because it is affected by the roast differently from the bread (because different things cause different sensations) and the impressions from one is a cause of a different desire from the impression

[8]Longuenesse (2017, p. 198, n.26) is referring here to Kant (1787B/1965, B414n.). Here is her entire quote: "Kant talks of 'degrees of consciousness' for what really seems to be, on his own account, different types of consciousness or discriminating capacity; one that is based on the mere quality of the sensation; the other that is based on a higher order act that calls on the unity of apperception....But the mere qualitative vivacity does not suffice for consciousness of the differences as such (such as the location of intuited objects in one space, or generation of concepts)."

[9]Kant is obviously a thinker within his own time in so far as he assumes the "lesser-ness" of animal cognition. Only in the most recent century (and sometimes grudgingly still) have the diverse cognitive capabilities of myriad animal species been accorded comparable standing with human cognitive capacities. Now, at least in some quarters, differences between humans and certain non-human animals, as well as variations in cognitions among the various species, are acknowledged and studied.

[10]Kant, I. (1762/1900). False subtlety of the four syllogistic figures. In AA2 [complete works in German], pp. 59–60.

of the other, according to the natural connection of his drives with his representations." But there are two rebuttals to Kant's view. First, this is true for human animals too, but perhaps not so obviously. We have concepts that can **seem** totally isolated from sensations, but are they not in fact derived from sensations more or less distantly? Second, the ability to discriminate perceptions/sensations on the basis of distinguishing those that could be significant from other sensations/perceptions, which can and should be ignored—this is itself a very basic and yet sophisticated conceptual capacity.

More important, and as Kant seems largely to agree, non-human animals also operate with a synthetizing unifying TUA, at least a form of such. And yet, there is indeed a difference. Accepting as given that for humans, the "**I think**" is the expression of the TUA, representing the TUA's successful function every time it is uttered, can we find an analog for non-human animals? I propose that we can: For non-human animals, each **agential action** expresses the successful function of their individual TUA, representing in the act's intention the synthesized coherence that characterizes the TUA for a singular subject. [Brakel's emphases throughout][11,12]

[11]For more on intentional acts and agency See Anscombe's (1957) seminal work, and my additions to it, in Brakel (2010, Chapter 4, pp. 91–133). I will say more on this topic in the current work in Chapter 3, in a subsection on the Agency View of *Self*.

[12]Tangential to Kant, but of importance for this project, there is another matter to consider regarding the conceptual capacities of the non-human animals discussed in Contrast Case 4. That is the fact that Trans-gender/Transsexual phenomena are not unique to humans. Some non-human animal species, particularly many species of teleost fish are capable of, and do switch either from male-to-female or female-to-male, depending on the species (see Francis 1992; and some preliminary comments in Chapter 1). As is true in general with the Contrast Cases presented in this project, the *Self* aspects involved in these non-human animal sex/gender transformations can be better understood than is the case for their human counterparts. This owes to the fact that for the non-humans, such switches are invariably contingent on initially cognizing the environmental conditions that obtain, and always toward increasing the reproductive selective fitness of the transitioning individual's *Self*. Note that at least one amphibian species too—the reed frog *Hyperolius viridiflavus. Copeia*—can switch from female to male (Francis 1992, p. 1). Much more about fish and sex change will be taken up both in Chapter 5, advancing a proper-function analysis of gender and Trans-gender, and in Chapter 3 on neuroscience brain research, which includes a section on sex-switching fish.

A different, almost opposite Contrast Case, Contrast Case 5—Depersonalization also is relevant in our consideration of Kant's Transcendental Unity of Apperception (TUA). It follows just below:

Contrast Case 5—Depersonalization
Contrast Case 5 concerns the phenomenon of depersonalization. Persons suffering from this symptom feel a heightened reflective consciousness, along with a troubling and persistent observationally detached stance toward their own physical bodies and movements thereof. It is as though they are Kantians in their experienced life—their TUAs make up the entirety of their notion of *Self.* The "I think" is the only part that seems real—they feel that they cannot know or feel their own bodies as real embodied aspects of self. Interestingly this heightened TUA can be considered almost a reciprocal to the non-human animal TUAs discussed in Contrast Case 4. There, for non-human animals, the TUA is a TUA of motivated intentional action rather than of thought. Returning to human considerations, Trans persons do not usually present with depersonalization—rather their bodies, rife with all the gender-at-birth empirical goings-on, seem all too real, as do their future hopes/plans regarding their gender transition in the future. But even were such a symptom volitionally available, it does not seem likely to improve the Trans person's life. Rather, it would produce a dis-embodied perhaps non-gendered life experience.

3.3 Implications of Kant's View of *Self* for the Trans Questions

The Kantian view offers no help in answering the three foundational questions about the *Self* arising with the Trans experience. Perhaps this owes to the fact that, as with the other views described, the Kantian view of *Self* too can provide no insight into any putative basic *Self* differences in Trans persons—differences that could potentially provide some traction. But, it remains the case that on Kant's account there are no discernible distinctions between the *Self* of Trans and Cis persons: Diachronic and synchronic unity of the singular, numerically identical, embodied, and subjective *Self,* as well as coherent representations of the objective outside world, and the subjective inner world, all function in the same way and to the same degree. Also, for all humans (Trans and Cis alike) the "I think" is the conscious expression of the TUA, which itself operates in conscious and unconscious mentation.

However, the contrast Kant puts forward between what he terms "representations to me" vs. "representations merely in me and therefore, nothing to me" does suggest something intriguing. Does the Trans person attempt to take certain representations and render them "merely in me and thus, nothing to me"? Here is a quote from Alic Custer-Shook, a Trans-gender man, before his transition: "It was February 2006, and I was 25 years old…I spent most of my free time researching how to transform my body from female to male. I could no longer look in the mirror because each time I was reminded of how uncomfortable I was in my body. Eventually, I covered up all the mirrors" (*New York Times* July 2, 2017 Sunday Styles Section, p. 11). This poignant quotation leads one to suspect that Trans persons, like Mr. Custer-Shook, might well have preferred to have had all of the natal-gendered perceptions, intuitions, and concepts as "representations merely in them and therefor nothing to them." But of course, and in a sense unfortunately, these representations, like so many others human representations, are under concepts, and therefore cohere as TUA mediated "representations to them." Would that there were some way to stop the automatically synthesizing TUA, because the representations from the natal-gendered *Self*, must feel not just coherent, but coherently diametrically wrong! And yet, despite their wrongness, they are "representations to these persons," not "merely in them;" quite far from being "nothing to them."

4 Freud

4.1 General View on *Self*

Freud (1923, SE 19, p. 26), in a work that is titled (in English) *The Ego and the Id*, famously characterizes the ego ("das Ich" translated as the "I" or the "self") as "…first and foremost a bodily ego." Filling out the role of the ego, and in a fashion agreed upon by modern psychoanalytic theorists, Freud describes the ego operating essentially as a collection of processes, whose functions are agential, aimed toward survival and real-world thriving. In this seminal work Freud puts it this way (p. 25):

> ...the ego [self] has been modified by the direct influence of the external world. Moreover, the ego seeks to bring the influence of the external world to bear upon the id [the seat of drives, impulsive strivings, and wishes] and its tendencies, and endeavours to substitute the reality principle [based on rationality and the assessment of veridically-aimed perceptions] for the pleasure principle [wish fulfillment-based] which reigns unrestrictedly in the id. For the ego, perception plays the part which in the id falls to instinct [drive and wishes]. The ego represents what may be called reason and common sense, in contrast to the id, which contains the passions.

Now returning to the importance of the body in constituting the ego, Freud is quite aware that the body serves in a unique and complex, multi-faceted fashion. Not only are the body's sense organs operative in registering perceptions of the external world, but the body itself is a part of the external world. The body has a great number of external manifestations perceptible to others as well as to oneself. Moreover, the body registers its own fluctuating internal states, and, to add one further fact: the external world clearly has various effects on bodily internal states. Paraphrasing Freud on these matters, Longuenesse (2017, p. 187) succinctly reports: "Now the information from the world comes via the body. The body is itself part of the world. So the organization of mental processes that constitutes the ego includes processes that represent states and changes of states of the body." And finally, Freud (1923, pp. 25–26), in a paragraph just prior to his assertion that the ego is first and foremost a bodily ego, holds:

> A person's own body...is a place from which both external and internal perceptions may spring. It is *seen* like any other object, but to the *touch* it yields two kinds of sensations, one of which may be equivalent to an internal perception. Pain, too, seems to play a part in the process, and the way in which we gain knowledge of our organs during painful illness

> is perhaps a model of the way by which we in general arrive at the idea of our body. [Freud's italics]

To complete this accounting of the ego and its functions, one must add that for Freud, it is the ego that is directly "in charge" of actions, at least under normal conditions and typical situations. He holds (1923, p. 25), "The functional importance of the ego is manifested in the fact that normally, control over approaches to motility devolves upon it."

So this is what we have so far: The ego is an agential entity with various processes, often termed ego-functions, that serve singularly and in concert: (a) to monitor inner goings-on from the id, including drives, and impulses—those that are biologically necessary for nutrition, mating, mate competition, prey and predator behavior—as well as desires and wishes, some of which are unacceptable, unattainable, viz. erotic and/or aggressive ones; (b) to register inner bodily states and bodily needs; (c) to represent the outside world (including aspects of one's body) realistically; (d) to mediate among these psychological and biological inner processes, in view of the existing external conditions, all toward assessing what is possible given both internal and external resources and constraints; and then (e) to organize id strivings, often modulating them with respect to determinations of beneficial vs. harmful actions. In sum, repeatedly and in an ongoing fashion in countless instances even within a single day and certainly over a lifetime, the ego operates (as a whole entity with its diverse and myriad ego-functions) to select and then initiate actions aimed at surviving and thriving in the outside world.

4.2 Problems with Freud's View

Freud's account of the *Self*, at least to the extent "the ego" is equated with the *Self*, is flawed in a way that can be usefully compared with the difficulties for the Humean *Self* discussed earlier. Given particularly his eliminativist-*Self* view, Hume should be asked these three questions: (1) Who is it, if not some overarching *Self*, that bundles the particular individual perceptions? and relatedly; (2) Who, if not some such overarching *Self*, does the looking for Hume's *Self*, when he is asked about it?; and

(3) Is that looker-for-the-*Self* and the bundler-of-perceptions, the same overarching *Self*?

Freud, with regard to his view of *Self*, has similar questions to address—questions that challenge his position. But first, for Freud there is a more basic matter to work out, namely: What is the nature of the Freudian *Self*—ontologically? The problem arises for Freud as his description of the ego (*Self*) renders it either (a) some sort of composite, or (b) something opposite to that, a homunculus-like, organizational, rationally-oriented, regulator, an agential meta-*Self*. In either case, Freud owes us explanations. Let's take the homunculus case first. Is the activity taken by this homuncular meta-self/ego identical with the *Self*? And if so, is the *Self* ontologically no more than this activity? If this is Freud's account of *Self*, it seems an active, processes-only counterpart to Hume's problematic passive contents-only, no-*Self* ontology. Freud's notion, however, is one step ahead of Hume's. To the extent that Freud is implicitly positing an overarching active organizational *Self*, he seems also to implicitly grant this *Self* some sort of (albeit uncharacterized) ontological standing.

Now if we take the other possibility, that the Freudian ego or *Self* is a composite, Freud's view is at least a step behind Hume's reductionist-*Self* as follows. Despite the difficulties with his account, Hume has at least detailed the nature of his "bundle of perceptions" along with the forces leading to the "fiction" of a coherent personal identity. Freud, on the other hand, has not specified which particulars, from which domains have a role in a Freudian account of *Self*. Nor has Freud suggested how the composite could cohere.

Perhaps we can help Freud out here. Although the "how" problem for composite coherence is much tougher, candidates for the particulars to be included are many. They could include id-derived drives, desires, wishes, and passions; prohibitions thereof (emanating from the "superego" to be discussed below); ego-mediated memories, thoughts, intentions, and plans; and finally, positive and negative identifications with other persons, these predicated on the ego and again the superego. Continuing along this line, the non-human animals of Contrast Case 4 might help lead the way in proposing particulars that would contribute to a more adequate, yet still genuinely Freudian composite type *Self*. Non-human animals and humans alike can be understood in terms of the interplay between biological (and psychological) id drives, and the agential ever ongoing assessments performed by the ego—monitoring reality, modulating drives, and initiating actions toward thriving.

Additionally, in humans, the ego also has an extra job—triangulating the id strivings and ego-mediated actions, along with the recommendations (sometimes prohibitions) from the conscience, known as the superego. Now it is a matter of some debate as to whether or not individuals of non-human species suffer the pangs of guilt and/or the more primitive superego punishments that humans routinely visit upon ourselves when we knowingly act in ways we ourselves view as bad or morally wrong. What is uncontroversial, however, is that aspects of superego contents do contribute to the human composite type *Self*—with varied responses to infractions large and small, differing also among different persons—but universally contributing. Now, having specified many particulars for the Freudian composite *Self*, another set of questions naturally arises: In addition to superego contents, does the superego also play a process role, actively functioning to cohere the composite's contents? If so, in what way(s)? Any answers to these questions might begin to address the still daunting problem: How are the composite parts of the Freudian *Self* coherently consolidated?

Pursuing this further, there is another human universal associated with the superego which not only warrants inclusion in a more complete Freudian account of the composite type human *Self*, but might also get us closer to specifying a cohesive force for consolidating the composite Freudian *Self*. Classified by Freud (1923) as belonging to a domain shared by the superego and the ego, a human's *Self* always includes "identifications" with other humans, both positive identifications—i.e., "I want to be like Person-A in X,Y,Z ways and I will do just that"—and the negative variety—"there is no way I'll end up being like Person-A, and especially not with respect of X,Y, Z." These identifications can be conscious, but are often unconscious. We find ourselves, for example, embodying the very traits we could not countenance in a parent; traits we swore we would never manifest. These identifications can be considered ego-mediated particulars. Yes, but as early post-Freudian psychoanalytic clinicians recognized, there is more—the superego too is involved, constructing an overarching type of cohesive identification called an "ego-ideal." Comprising many different traits, one's ego-ideal contains aspects of persons to which one aspires. The ego-ideal represents what/who the *Self* would like to be like. As such it has elements of the two possibilities described for the Freudian *Self*—the overarching meta-*Self*, which in this case would actively mark and select the "desirable" traits; and the composite *Self*, constituted by the contents associated with these very traits.

The above notwithstanding, the ego-ideal is not free from the problems inherent in each of the two Freudian-*Self* accounts. First, as stated just above, there are unconscious identifications far different from those consciously desired. This challenges the notion that the ego-ideal forms a composite *Self* based upon traits and identifications deliberately selected and cohering as "desirable." Second, and more problematic, the meta-*Self* nature of the ego-ideal and its operations, other than it being active, remains mysteriously homuncular and ontologically opaque. Nevertheless, despite these issues, a revised Freudian account of *Self*, one endorsing both the meta-*Self* activity and the composite-*Self* contents, might actually represent at least a small advance.

4.3 Implications of Freud's View of *Self* for the Trans Derived Questions

Can the original or even the revised Freudian notions of *Self* answer, address, or even accommodate the foundational questions occasioned by the Trans-gender phenomenon and Trans-gendered selves? Consider the Freudian dictum that "the ego is first and foremost a body ego." For Trans persons, there seems nothing unusual here as the dictum is lived in the usual way. Normal bodily sensations, drives, and impulses deriving from inside one's body, as well as perceptions of one's body—one's own and that of others—all seem to register indistinguishably from those processes in persons who remain lifelong Cis persons. And yet at some point in development, perhaps different for different Trans persons, the very realistic perceptions of the Trans person's own body, while recognized as belonging to this person's own body and own *Self*, feel almost exquisitely painful and definitely wrong. Again, I offer the same quote as above from Alic Custer-Shook, a Trans-gendered man, spoken before his transition, as it demonstrates with heartfelt clarity this tripartite phenomenon—it is my body, but it is the wrong body, and I feel pain: "It was February 2006, and I was 25 years old…I spent most of my free time researching how to transform my body from female to male. I could no longer look in the mirror because each time I was reminded of how uncomfortable I was in my body. Eventually, I covered up all the mirrors" (*New York Times* July 2, 2017, Sunday Styles section, p. 11). The quote is the same as before. This is to demonstrate that while Alic Custer-Shook considered the female body in the mirror "wrong," she/he still recognized this body as his/her *Self* and uniquely so.

So far then, it seems that the Freudian view of *Self* cannot provide a better understanding for the foundational questions that arise about the *Self* given the Trans experience. But what if we focus specifically on the ego-ideal, both in its selective processing, and in its resultant contents? Can this addition to the Freudian *Self* allow us more room in which to adequately solve the puzzles about the *Self* concept brought to light by the Trans phenomenon? For instance, we can wonder: Do Trans persons experience what I will term "ego-ideal switch-ups" of a certain kind? Here is what I have in mind: For Trans persons (at least for a subset thereof), could it be that all the traits admired and selected, and moreover all the persons chosen as worthy of emulation, are represented only by people of, and traits associated with the non-natal gender?

Maybe this could be a possibility. And yet, at most this gives us a preliminary, and rather flimsy terminological handle, not real traction. This is the case because, currently at least (and at last), it is mostly true that the great majority of traits can be well accommodated by either gender. Furthermore, people can readily identify with a person, or even exclusively with persons of their non-natal gender, without knowing or believing or feeling that one is of that other, and therefore without any desire to transition to the non-natal, other gender. Thus, Cis-gendered persons describe knowing their gender effortlessly, each perception, over years and decades—even for those people who are not particularly gender-standard—somehow fitting and congruent with the gendered assigned at birth. So called "Tomboys" are a good, if rather conventional example. They are girls, who know they are girls, even while preferentially engaging in activities standardly thought of as masculine and in the domain of males.

Thus, as is true of the other classic views, the Freudian account of *Self*, even when expanded, has not helped answer the three foundational philosophical questions occasioned by the Trans phenomenon.

* * *

Before we move on to investigate Modern Philosophical views of the *Self* in Chapter 3, let's take up a psychiatry related Contrast Case–Contrast Case 6—Hallucinations and Delustions.

Contrast Case 6—Hallucination and Delusions

Following our discussion of Freud—whose views on *Self*, while not without difficulties, revolutionized psychiatry—it seems apt to next consider the psychiatric symptoms of hallucinations and delusions as constituting Contrast Case 6. The contents of hallucinations and delusions are often experienced as alien to the *Self*. Yet, no matter how external these contents feel, they are really self-contents. People with chronic psychotic illnesses learn to regard their hallucinatory and delusionary contents as derived from *Self* notwithstanding their alien feel, while at the same time they practice ignoring these contents in the service of continued behavior in the normal, expectable, acceptable range.

Although I have heard that some psychoanalysts and psychiatrists regard the "wrong gender" knowledge of Trans persons as delusional,[13] this is in my view a serious error. Trans persons, particularly those upon which this project is focused, are stipulated as largely mentally healthy, suffering from no psychotic illness, nor indeed any major psychiatric diagnosis. Thus, the Trans-gender feelings, beliefs, desires, and longings of these people, as well as any imagined future experiences, are neither delusional nor hallucinatory. Similarly, their natal gender perceptions,

[13] For example, Maria Ines Carneiro, a Brazilian psychoanalyst, has in a personal communication, made this claim. She also has expressed this view in Carneiro (2018).

sensations, and experiences also are neither delusional nor hallucinatory; this although they are more likely to be felt as alien. Nonetheless, despite the alien feelings occasioned, these natal gender contents are still acknowledged as part of the *Self* and real—all too real.

* * *

To recap: The Freudian account of *Self* gets us no farther in answering the central epistemological, ontological, and conceptual questions: For a Trans person, how does one know one's natal gender is "wrong?" What ontological aspects of the *Self* are involved? And what or how should the very concept of *Self* be expanded or changed in order to derive a more satisfactory understanding of these fundamental philosophical questions brought to light by the Trans phenomena?

Indeed, none of the four seminal and historically important accounts of *Self* examined here—not that of Descartes, Hume, Kant, or Freud—has proved sufficient or even terribly helpful in explaining the aspects of *Self* that would occasion these questions for Trans individuals, and not for Cis persons. And it follows that none of these four classic *Self* views advanced our quest to answer the foundational epistemological, ontological, and conceptual questions about *Self* that Trans phenomena bring to the fore. So, let us try, in Chapter 3—Modern Philosophical Views of *Self*—to get at this from another angle.

Bibliography

Anscombe, G. E. M. (1957). *Intention.* Cambridge, MA: Harvard University Press.

Brakel, L. A. W. (2010). *Unconscious knowing and other essays in psycho-philosophical analysis.* Oxford: Oxford University Press.

Brakel, L. A. W. (2013). *The ontology of psychology: Questioning foundations in the philosophy of mind.* New York: Routledge.

Carmeiro, M. (2018). *Concerning cures, transgenderalities, and misconceptions.* Unpublished manuscript.

Chisholm R. (1969/1994). *On the observability of self: Philosophy and Phenomenological Research* (Vol. 30, pp. 7–21) (Reprinted in *Self Knowledge*, Chapter 5, pp. 94–108, Q. Cassam, Ed. Oxford: Oxford University Press).
Coplston, F. (1963). *A history of philosophy: Volume 4: Modern philosophy: Descartes to Leibniz*. Garden City, NY: Doubleday Image Books.
Descartes, R. (1641/1969). Meditations on first philosophy. In E. Haldane & G. Ross (Eds. & Trans.), *The philosophical works of descartes* (Vol. I, pp. 131–200). Cambridge: Cambridge University Press.
Edelman, G. (2003). Naturalizing consciousness: A theoretical framework. *Proceedings of the National Academy of Sciences, 100,* 5520–5524.
Francis, R. (1992). Sexual lability in teleosts: Developmental factors. *The Quarterly Review of Biology, 6,* 1–18.
Freud, S. (1923). The ego and the Id. In J. Strachey (Ed. & Trans.), *The Standard Edition* (Vol. 19, pp. 3–66). London: Hogarth Press, 1961.
Guyer, P. (1992). Transcendental deduction of the categories. In P. Guyer (Ed.), *The Cambridge Companion to Kant* (pp. 123–160). Cambridge: Cambridge University Press.
Hume, D. (1739/1978). *A treatise on human nature*. Book I/Part IV/Section VI (L. A. Selby-Bigge, Ed.). Oxford: Oxford University Press.
Kant, I. (1781A/1787B/1965). *Critique of pure reason* (N. K. Smith, Ed.). New York: St. Martin's Press.
Longuenesse, B. (2017). *I, me, mine: Back to Kant, and back again*. Oxford: Oxford University Press.
Malcolm, N. (1965). Descartes' proof that his essence is thinking. *Philosophical Review, 74,* 315–338.
Penelhum, T. (1955/1966). Hume on personal identity. *The Philosophical Review, 64,* 571–589 (Reprinted in *Hume: A Collection of Critical Essays*, pp. 213–239, V. C. Chappell, Ed. Garden City, NY: Doubleday Anchor Books).
Pike, N. (1967). Hume's bundle theory of the self: A limited defense. *American Philosophical Quarterly, 4,* 159–165.
Wolff, R. (1960/1966). Hume's theory of mental activity. *Philosophical Review, 69*(3), 289–310 (Reprinted in *Hume: A Collection of Critical Essays*, pp. 99–128, V. C. Chappell, Ed. Garden City, NY: Doubleday Anchor Books).

3
Modern Philosophical Views of *Self*

Abstract The three foundational philosophical *Self* questions are again the focus. Modern views investigated include many and varied accounts of (1) *Self*-Constitution; and (2) the *Self* that one is concerned with when one is concerned with one's survival. Under each of these two categories there are a few views that feature evolutionary considerations—Agency views, for example, and an account owing to S. Buss highlighting the *Self* with respect to constraints of minimal human flourishing. This chapter also explores a number of Classic Thought Experiments designed to reveal the nature of *Self*—Body-Brain Exchange (B. Williams); Closest Continuer (Nozick); Tele-transporting with Percentage change (Parfit)—experiments which will be adapted for empirical Experimental Philosophy studies in Chapter 7.

Keywords Modern *Self* views · *Self*-Constitution accounts · *Self* I care about re self-survival · Classic *Self* Thought Experiments · Minimal human flourishing *Self* view

Unsurprisingly, there is a huge modern literature on this ever-intriguing subject—the *Self*. First, in a section centering on the *Self* in self-constitution, four categories are discussed, within which are several

L. A. W. Brakel, *Investigations into the Trans Self and Moore's Paradox*,
https://doi.org/10.1007/978-3-030-44645-1_3

accounts. Then, in a second subsection, we will investigate what we can learn of the *Self* when we ask "What is the *Self*/the 'me' I care about, when I care about my survival?" Five different types of answers to that "me" question are taken up.

1 The *Self* in *Self*-Constitution

1.1 Pruning Views of *Self* vs. Non-*Self*

Even within this "Pruning" category there are several different views. Although each is distinct, and many interesting, the focus here will be on the essential common feature of three modern "Pruning"-*Self* prototypes. As the name implies, these accounts rely on a person's capacity to select certain contents, traits, and aspects as constituting the *Self*, while "pruning" and thereby rejecting other contents, traits, and aspects. The three Pruning views we will consider are the Endorsed-*Self*, the Rational-*Self*, and the True-*Self*.[1]

Endorsed-*Self* Views

Perhaps most prominent among the Pruning views is the Endorsed-*Self*, with the clearest (and most well-known) account found in the early writings of Harry Frankfurt. For Frankfurt, in order to achieve selfhood, each person must actively constitute his/her own "will." He describes how this is done in a series of essays entitled The Importance of What We Care About. Frankfurt (1971/1988) begins by comparing first-order desires with second-order desires (p. 16). "Someone has a desire of the second-order either when he wants simply to have a certain desire or when he

[1]There are classical "pruning" views too. These include those of Plato, Kant, and Joseph Butler, as well as contemporary-Kantians such as Tamar Schapiro and Christine Korsgaard. (See the Bibliography for particularly relevant works.) Although Schapiro's article and the three classic accounts will not be addressed in this project, portions of Korsgaard's work will be briefly discussed below in the section on Rational-*Self*. The common feature among all of these accounts, classic, contemporary-Kantian, and modern alike, is actually the "pruning" process itself: Identifying the *Self*-proper as distinct from aspects that are non-*Self*, and then pruning away those non-*Self* components.

wants a certain desire to be his will." He continues (p. 16): "[T]he latter kind, I shall call his 'second-order volitions'...[and] it is having second-order volitions... that I regard as essential to being a person." He explains this by contrasting the willing with the unwilling addict. The unwilling addict has a first-order desire for the drug and, unlike the willing addict, also a conflicting first-order to desire to not succumb to his addiction. In addition, Frankfurt maintains that (1971/1988, p. 18): "The unwilling addict identifies himself...through the formation of a second-order volition, with one rather than the other of his conflicting first-order desires. He makes one of them more truly his own and, in so doing, withdraws himself from the other." The upshot, for Frankfurt (p. 18), is that the unwilling addict can rightfully claim that "...the force moving him to take the drug is a force other than his own." Only these second-order volitions, with which one actively identifies and thereby endorses, are for Frankfurt, internally located, essential aspects of a person, a *Self.* What about the unendorsed, withdrawn-from, first-order desires? By implication they have been "pruned" and are not central essential aspects of the person's *Self.*

Gary Watson (1975) raises an important objection. Watson (pp. 217–219) claims that Frankfurtian second-order desires cannot do the work Frankfurt sets out, because second-order desires, being merely other desires, have no special status. In addition, as Frankfurt himself acknowledges, one could raise an infinite regress objection here and ask: Why stop at the second-order level? Why not form third-order desires with regard to our second-order desires? Partly in response to objections like this, Frankfurt has modified his view. In subsequent work, he holds that the infinite regress can be stopped and the possible arbitrariness of one's second order desires countered, by a "decisive commitment" to a certain course of action. The "decisive commitment" adds an intentional potency not necessarily present in a desire, even one of considerable intensity.[2]

[2]See particularly Franfurt's (1992) The faintest passion. This is both an address and an article. (See Bibliography for full citation.)

Rational-*Self*

Christine Korsgaard (1999) disagrees with Frankfurt that endorsed second-order desires, even those buttressed with Frankfurt's firm commitment amendment, could be central in constituting one's *Self*. For Korsgaard, a staunch proponent of the Rational-*Self*, only reason can do the legislative work needed for self-integration—work that she deems necessary for self-constitution. Thus, embracing Platonic harmony of the soul as figuring importantly in any viable model of self-constitution, Korsgaard (1999, p. 12) writes: "Appetite presents the proposal; reason decides whether to act or not, and the decision takes the form of a legislative act."

Moreover, Korsgarrd's view, as with other Rational-*Self* views, entails pruning; and pruning toward reason and rationality is an emblematic, easy-to-discern type of pruning. We (Brakel & Fileva, unpublished ms, p. 3) put it this way: "…an agent, on the pruning view, can achieve inner harmony by giving affirmation to the central aspects of his or her self, while refusing to license the undesirable parts…[And]…on pruning view[s], the self-proper is to be identified with some 'higher' agential component of a person's mental economy, paradigmatically, reason and rationality…[with] reflection understood as the process of normatively assessing one's first-order attitudes in light of reasons."

True-*Self* Views

This category is heterogenous. There are many different authors contributing views counting as category members, with all such views pointing toward a *Self* that is classified as the Real/Authentic-*Self*. The social psychologist Nina Strohminger and colleagues (2017, p. 552) in an article titled "The True Self," list many such members, and comment as follows:

> Sometimes it is called the real self (Koole & Kuhl, 2003; Masterson, 1988; Rogers, 1961; Sloan, 2007; Turner, 1976), the ideal self (Chodorkoff, 1954; Higgins, 1987; D. T. Kenny, 1956), the authentic self (Cable, Gino, & Staats, 2013; Johnson & Boyd, 1995), the intrinsic self (Arndt,

> Schimel, Greenberg, & Pyszczynski,2002; Schimel, Arndt, Pyszczynski, & Greenberg, 2001), the essential self (Strohminger & Nichols, 2014), or the deep self (Sripada, 2010).[3] However, a consensus is emerging over the term 'true self'...[a term the authors themselves adopt, citing also] (Bargh, McKenna, & Fitzsimons,2002; Johnson, Robinson, & Mitchell, 2004; Landau et al., 2011; Newman, Lockhart, & Keil, 2010; Schlegel, Hicks,Arndt, & King, 2009; Sheldon, Ryan, Rawsthorne, & Ilardi, 1997).

Following Strohminger's lead, I will take the term True-*Self* to represent the many different accounts—philosophical and psychological—which despite their range, all hold together as a class. This owes to two important interrelated reasons. First, they are all Pruning-*Self* views; and second, they all stand in contradistinction to the "Superficial or External-*Self*," aspects of which, each of these accounts suggest, can and should be pruned.

A natural question arises: What are the features that constitute the True-*Self*? One's moral character traits seem central. Indeed, according to a recent study examining factors potentially disrupting to one's sense of personal identity, Chen et al. (2016) found that moral traits are regarded as most centrally causal, and most deeply rooted. Moreover, this assessment skews toward the "good" side of morality, and relatedly favors reason and rational control over emotionality. Thus, when Person-A lashes out during an argument, Person-B loses the resolve to stop smoking, and Person-C impulsively purchases an expensive house for a new lover, the standard notion might be expressed in this way: "Person-A has been overcome with anger; Person-B has given in to an addictive craving; and Person-C is controlled by a strong emotion." These typical phrases illustrate the view that the True-*Self* in each of these cases, has been supplanted, and it is hoped temporarily so, by these external and superficial self-aspects. Beyond this, were someone to claim of Person-A, "A is not himself," this would be an even stronger expression of the contrast between external and True-*Self* features. The pruning then will

[3] Sripida's deep self view will be briefly taken up later in the "Cares" category of *Self*-views. The other accounts are merely listed, and not discussed further.

attempt to rid the essential True-*Self* of these not-good, non-rational, highly emotional features.

Problems with Pruning Views of *Self*

There is a serious two-pronged problem with all three of the Pruning-*Self* views discussed. First, there is the implication of an almost omniscient, overarching Knowing-*Self*,[4] capable of picking out the most "authentic" parts of that same *Self*—parts, which just happen to be the most rational, moral, and endorsable *Self* aspects. Second, this Knowing-*Self*, also has access to the Pruner-*Self* which functions (in a not quite omnipotent way) to extrude the less rational, more impulsive, less moral parts; in other words, to prune away those less than desirable aspects from the *Self*. How might one do this? Alas, this would be a neat trick if we humans could indeed construct ourselves in this manner; but as it seems pretty clear that we cannot, perhaps these Pruning-*Self* accounts might better be considered Aspirational-*Self* or Wished-for-*Self* views. (More snarky critics might even call them Wishful Thinking-*Self* views.)

That we cannot self-constitute in the manner the Pruners outline evokes a variety of critical comments among moral philosophers. Brakel and Fileva (unpublished ms, p. 8) summarize some of these criticisms, adding our own: Thus, we hold that the Pruners "…accord too great a role to reason and second-order attitudes and too little to conative elements and first-order attitudes, such that they overemphasize the conscious, active role we play…and fail to recognize the importance of seemingly passive or even unconscious aspects of the self—those desires, cares, and character traits, weak-will, lapses of attention, etc., neither consciously chosen nor endorsed…" (As a preview, note that some opponent accounts to the Pruning-*Self* will be taken up in a section just below. A bit later, a few Conative-*Self* views will be discussed).

[4]There might exist a resemblance of some kind between this sort of Knowing-*Self* and the *Self* who knows that the natal gender is "wrong" in Trans persons. This similarity could prove useful if ontological and epistemological aspects of the Knowing-*Self* could be understood. But alas, this is unlikely as is elaborated in the text to follow.

Implications of Pruning Views of *Self* for the Trans Occasioned Questions

The Pruner-*Self* is constituted in part by pruning away certain behaviors, traits, and habits on the recommendation and activity of a Knowing-*Self* that functions to "know" which parts are to be valued and kept and which to be extruded. This "knowing" part of the *Self* uses particular markers to separate that which needs pruning from that which should be kept. The more rational, moral, and endorsable elements are to be kept as essential and internal parts of the *Self*, whereas the less rational, impulsive, emotional, and addictive are tagged as external, ready for removal.

Now can this help us understand how and what Trans people know, when they know they are the "wrong" gender? The most immediate answer is: No, any marker for gender seems quite orthogonal to the positive value-based assessments that the Pruner Knowing-*Self* utilizes. But this answer is incomplete. While gender is generally value-free (ideally), a Trans person might well be inclined to tag-for-excluding aspects of the natal, i.e. "wrong" gender. And yet, there is still an internal/external problem. Could young GB, our emblematic Trans person, take all of the internally mediated female bodily sensations, perceptions, impressions, and registrations accrued, and render them paradoxically external? Should GB do this? Furthermore, and simultaneously, could (and should) GB instead whole-heartedly endorse as internal some version of male bodily mediated sensations, impressions, etc. as the deep and true self? Meanwhile, should all of these internal/external exchanges begin while transitioning is just imagined, continuing more intensely as transitioning becomes actually contemplated? These intertwined tasks seem not only daunting, but so basic and disorienting as to be potentially damaging, rather than enhancing of GB's *Self*.

Even if the "wrong" gendered contents could be tagged and cleaved away, the original ontological question animating this project remains vexing. In addition, we now have this question in three versions. (1) For Hume (as was discussed above): To what *Self* or aspect thereof, do the impressions/perceptions report, such that Hume's only "fictional" self can emerge? (2) For those holding a Pruning-*Self* view: What is the ontological nature of the knowing-*Self* who knows and marks the essential aspects of *Self* as keepers, while tagging other traits to remove? (3) For the Trans person: What *Self* or aspect thereof, realizes and deems that this person's own traits, behaviors, habits, impressions/perceptions are those of a *Self* that is differently gendered from the natally assigned "wrong" one? And now, another question: Can these parallel prove useful?

Perhaps the parallels do suggest some progress, at least in terms of comprehending the ontological question. But, on the equally troubling epistemological question, the Pruning views of *Self*, provide no parallels. Thus, whereas on most (if not all) of the Pruning views, one can grasp how one might select "better" (i.e., more rational, reason-based, and moral) elements to keep as *Self* and "worse" ones to extrude; it is a total mystery as to how a Trans person can know that the natally-gendered elements are "wrong," designate them as such, and feel their "wrongness" profoundly, and whole-heartedly.

As the pruning views of *Self* have, like the other views explored so far, proven of but very limited success in aiding our quest to answer foundational questions about the *Self* arising in the context of Trans persons, let's turn next to accounts in sharp disagreement with Pruner-*Self* views.

1.2 Opponent-to-Pruning Views of *Self* vs. Non-*Self*

In a view that might best be characterized simply as an Opponent-to-Pruner-*Self*, Terrence Penelhum (1971, p. 671) starkly states: "That part of us from which we wish to dissociate ourselves is as much a part of us as that which we wish to identify." Penelhum (p. 672) continues: "...that part of himself, past or present, with which he does not identify is, tiresomely, just as much part of him as that which he does." He concludes (p. 674): "We would all reject as evasive fantasy the suggestion that the desires that moved a man to act were not really a part of him; and...we can correctly say that these desires have to belong to *someone* [Penelhum's italics], and must necessarily belong to the person who also owns the body that performs the actions…"

Building on criticisms such as these, we (Brakel & Fileva, unpublished ms) have advanced a view of self-constitution we term "Understanding First." On this view, an agent recognizes, owns, and engages with undesired and undesirable parts of the self, admitting that these aspects belong to the real-*Self* no less than the endorsed, rational, and moral traits. The engagement entails a non-judgmental investigation of how and why these less than ideal self-aspects developed, allowing a reassessment and perhaps a realignment. Our account, like Penelhum's, is an Opposite-to-Pruning-*Self* view. However, after the necessary step of enlarging and expanding the *Self* in order to accommodate the negative elements, and after serious but compassionate self-understanding of their meanings and origins, there is the potential for salutary reconsideration.[5]

Problems for the Understanding First View of *Self* and Implications for the Trans Derived Questions

No better than the other views, and perhaps worse, our Understanding First view of self-constitution devolves very quickly into incoherence for Trans persons. Consider GB. Before any transition GB should do what? Embrace "her" alienated femaleness along with "his" wishful maleness? Embrace her/his notions of felt-maleness and the femaleness that he/she regards as external? Extrude something? But when? How long should GB live with enormous conflictual aspects of *Self*?

1.3 Conative Views of *Self*

Partly in response to problems with the Pruning-*Self*, and also relevant to the problems with opponent views like "Understanding First," several

[5]A psychoanalytic treatment is an excellent way, albeit not the only way, to instantiate the Understanding First View of *Self*.

Self accounts have developed featuring conative aspects as the core elements in the self-constitution project. These Conative-*Self* views can be divided into two sub categories—the Integrated-*Self* and Central Cares-*Self*.[6]

The Integrated-*Self*[7]

Taking the work of Arpaly and Shroeder (1999) on "Praise, Blame, and The Whole Self" as representative of Integrated-*Self* conative accounts, the authors (p. 172) explain that while their Whole-*Self* view privileges some aspects of *Self* as central over other aspects, they reject the division between the rational and emotional, between reason and desire: "Rather there is a continuum…some fitting well with the agent's character—which includes both her Reason and Appetite—and some fitting poorly; of better- and worse-integrated motives." Clearly the task of the Whole-*Self* theorist is to provide some guidelines as to what counts as "well integrated," and these authors (pp. 172–173) aver to beliefs and desires, rather than character traits, as the grounds on which this can be accomplished. They contend (p. 173): "So, beliefs and desires produce actions. We will say that beliefs and desires are well-integrated within a person to the extent that (1) they are deep; and (2) they do not *oppose* other deep beliefs or desires." They continue (p. 173), "A belief or desire is *deep* insofar as it is a powerful force in determining the actor's behavior, deeply held, deep-rooted. Deep beliefs tend to resist revision…and deep desires tend to be satisfied with preference over shallower desires…where a choice is forced [Arply and Shroeder's emphases throughout]."

---

[6]In addition to the three accounts to be discussed in the subsections on the Conative *Self* views, see also the works of Nomy Arpaly (2003) and Thomas Scanlon (1999). *Unprincipled Virtue: An Inquiry into Moral Agency* (Oxford: Oxford University Press, 2003); and Thomas Scanlon, *What We Owe Each Other* (Cambridge, MA: Harvard University Press, 1999).

[7]Before discussing this category, it should be noted that self-integration figures in other categories of Self-constitution too, albeit not as the central feature. This can be seen particularly in the above discussion of Korsgaard's Rational-*Self* view, and in a variety of forms in the Central Cares-*Self* views which will be presented in the next subsection below.

The Central Cares Account of *Self*

Chandra Sripida (2016) offers an integrated and conative account of *Self* constitution predicated on what Sripida terms a person's central cares. He describes and explicates his notion (p. 1206): "I propose a *conative* view that says a person's deep self consists of her *cares.* Cares are a distinctive kind of mental state set apart from other mental states (e.g., beliefs and ordinary desires) by their unique functional role. The distinctive syndrome of dispositions associated with cares…makes them well suited for constituting a person's deep self" (pp. 1208–1210). Sripida outlines these in four parts. First,

> …caring about something always serves as a source of *intrinsic* [distinct from instrumental] motivation for actions that promote the achievement of that thing …Second, caring is associated with a distinctive set of *commitmental* effects. In caring about X, the person not only is intrinsically motivated to bring about X, the person is *in addition* intrinsically motivated to continue caring about X… Third, caring is connected to evaluative judgment; a person who cares about something is disposed to form judgments about …it in a normatively favorable light…Finally, caring is also associated with a rich and distinctive profile of emotional responses. [Sripida's emphases throughout]

Summing up, Sripida holds (p. 1211) that caring (in the ways described), reveals what is genuinely important to a person, contributing to a person's "fundamental conative point of view," and constituting one's deep self.

Sripida makes two additional useful claims, both concerning integration: First he (pp. 1211–1212) distinguishes between (a) one's ontological deep self—the deep self that actually reflects a person's central cares; vs. (b) one's psychological deep self—that which a person takes to be his/her deep self. This distinction not only suggests that a person may demonstrate a coherent self-integration beyond his/her own awareness, but also introduces the idea that someone can be mistaken about the nature of one's deep-*Self.* Of importance, this distinction marks Sripida's view as offering an improvement over what can amount to an integrated, but false, "wished for-*Self*" arising after pruning. Second, Sripida argues that the deep-*Self*, constituted by central cares, can nonetheless be a conflicted self, and in this sense not perfectly integrated. He explains (p. 1226): "on the mosaic conception that I favor, deep selves are potentially complex, heterogeneous things…To care for X, believe Y is incompatible with X, and care for Y is not irrational. Rather, it is the human predicament…"

David Shoemaker (2003) offers another variant of a Central Cares-*Self*. Stressing (p. 105) that "None of one's current cares are under one's control at the current time…[in that] we cannot simply start or stop caring about something…" Shoemaker maintains that it is our very cares that shape our will; this quite opposite to the view that we consciously endorse one set of cares over another, then willing that set to be representative of our *Self*. Indeed, Shoemaker avers (p. 108): "This is not to suggest…that having our cares determine our decisions is akin to having them imposed on us by some outside element. Instead, the determining forces are internal to, and (what is key) representative of, us." Shoemaker then takes this idea one important step further (p. 112): "Identification occurs passively, garnering its authority for self-determination from one's nexus of cares." Moreover, he holds that although conflicting cares exist, along with desires to endorse the more ideal ones, Shoemaker's position is that in the nexus of cares, the stronger ones will win out (p. 115). Put simply, Shoemaker's core claim is (p. 113): "…our identity as functioning, well-developed agents…is constituted by our nexus of cares." Thus, when we ask, "Who are we?" we can answer, "We are what we care about."

Problems with Conative Views of *Self*

Conative views of the *Self* and self-constitution certainly do not have the selectivity problems of pruning accounts. As such, they are quite appealing. But conative views do have another set of problems. Since conativists suggest that the constitution of one's deep-*Self* includes unendorsed, passive, and even objectionable parts, no less than the "good," i.e., moral, and rational elements embraced whole heartedly, serious inner conflict is allowed, almost promoted. Moreover, there is a seemingly definitional circularity in picking out what should be considered "integrated" and reflective of the deep-*Self*, and a related regress problem. This can be seen most clearly in the Central Cares-*Self* accounts. What we deeply care about determines our deep-*Self*, i.e., who we are. What we deeply care about shapes our wills and guides our actions. But how do we get to care about what we care about? And, what is the nature of the *Self* (or aspects thereof) that cares about what we care about, and cares about those cares deeply?

Implications of Conative Views of *Self* for the Trans Questions

The two questions raised just above are recognizable versions of the initial ontological and epistemological questions motivating this project. The Trans persons with whom we are concerned here—those who "know" their natally-assigned bodily-predicated gender is the "wrong" gender—are persons for whom this contrast has to be a source of both central care and intense conflict.

On the Integrated-*Self* account, the ontological question is very much at issue in so far as this view would predict that Trans persons could not form a coherent integrated-*Self*. Yet, there is much evidence that Trans persons, even those with dysphoria about this essential conflict, have quite coherent, stable integrated selves, much as do Cis persons.[8] Who, what *Self* (or part thereof), can do that integrating work?

With regard to the Central Cares views of *Self*, it is the epistemological how question, which not only continues unremittingly, but now expands in a troubling manner. Take the original question: How do Trans persons know their body is the "wrong" gender? For Trans persons—those contemplating transition, those intending to transition, and those who have transitioned—this question must be linked to the "wrongness" itself being among the most (if not the most) central of cares. But how has the "wrongness" happened?

[8]See for instance, Roy Grinker's *New York Times* opinion page piece (December 7, 2018), titled, "Being Trans is Not a Mental Condition." See also Chapter 1, footnote 5 for more on Grinker's essay and related material.

* * *

So where are we now? So far, we must conclude that our quest to find a philosophical view of *Self* that can offer answers to epistemological and ontological questions arising when considering the Trans phenomena has been disappointing. Moreover, as the epistemological and ontological questions remain, so too the conceptual matter regarding the *Self* must continue unsettled. So, let us try to get at this from yet another perspective, again with the hopeful hypothesis that some insight with respect to the foundational questions about *Self* occasioned by the Trans phenomenon, can aid in advancing a more complete philosophical view of the nature of *Self*. One new creative angle, owes to an article by Sarah Buss, whose account is briefly outlined just below.

1.4 *Self* Constrained by Minimal Normal Human Flourishing

Sarah Buss (2012) adds a vital element to the understanding of the *Self*. She does this with regard to intentions, actions, and behaviors, all of which must be consistent with what she terms "the minimal human flourishing condition." She states (p. 659): "In determining the boundaries of the self that do not simply presuppose that a person can be identified with her character or personality (or with any relatively stable psychological disposition), the human flourishing condition also determines the boundaries of a person's character and personality." That this is a biologically relevant species-specific condition is explained by Buss in a seminal paragraph (p. 660):

> I believe that the human flourishing condition is the key to the unique respect in which autonomous agents play a determining role in their actions: a human being fails to act autonomously if and only if she forms her intention under … conditions that are elements or symptoms of human malfunctioning…[There is an] important sense in which an agent's identity depends on the kind of agent she is, and the important

> sense in which the necessary conditions of *self*-determination [Buss's italics] are thus normative. To be a human agent is to be a representative member of a species.

For Buss then (p. 667), a person cannot operate as an autonomous agential human *Self* when pain, fear, great disability, or symptoms of pathology "…are extreme enough or exceptional in some other way that renders them at odds with minimal human flourishing…"

Implications of Minimal Human Flourishing *Self* for the Trans Questions

Buss's view, as is true for all the others taken up so far, cannot provide answers to the ontological or epistemological *Self* questions central to understanding the Trans experience. However, the species-specific, biologically relevant condition she suggests does point the way to a consideration of an evolutionary based proper-function account for gender aspects of the *Self* in Trans persons, both as is different from, but also similar to, such an account for Cis people. (Just such a proper-function account of gender will be offered in Chapter 5.)

2 The *Self*—The "Me" I Care About When I Care About My Survival[9]

The next category of modern philosophical *Self* accounts considers the *Self* in terms of the following straightforward response to a most basic question: What is the *Self*? It is the "me" I care about when I care about my survival. Within this category a variety of different ideas of that "me" will be explored.

[9] Much of this section is revised material from a chapter first published in Brakel (2010, Chapter 4, pp. 91–133). "Agency—'me'-ness in action."

2.1 Human Fission

Amoeba reproductive division presents the tantalizing reality of no-death. The parent amoeba does not die, but does divide into two progeny cells, each genetically identical both to its sibling cells and the original parent. The two progeny cells also do not die, they survive as two until they each divide.[10] The amoeba situation lends itself to classic Thought Experiments, contemplating human fission in order to reveal answers to the nature of the *Self*—the "me"—I'm concerned with, when I'm concerned with my own survival. There are several such experiments, each with a number of variations on this theme of human fission.[11]

One version of the basic story is as follows: You are about to die of a cardiac arrest. Your brain however is in good shape and can be transplanted, and there are three possibilities. (1) Your brain is transplanted into a clone of you, prepared just for this eventuality; (2) Your brain will be transplanted into a non-carbon-based replica of you, made ready as above; (3) Your brain will be transplanted into another person's body. The body is in good shape, as the person suffered a sudden but discrete brain death. Moreover, in this third case the body resembles yours. Note that since this last case can be made more interesting if the other person happens to be your identical twin, we will go with this!

Now for some variations: The first one concerns the opening setup. Now, in this new version, you have been about to die, but you do not. The surgeon still believes that soon you will have the predicted cardiac death, but she can no longer be so sure. Since humans can thrive with only half of a brain, the surgeon removes for transplant only one half

[10]For amoebas there are actually two types of reproduction. One is the mitotic division described in the text, in which two genetically identical diploid daughter cells are formed, each also genetically identical to the parent cell. The other form of reproduction is meiotic division. Here the paired chromosomes are split before their genetic material is doubled, forming a pair of haploid daughter cells. Only then is the genetic material doubled with another division occurring after some reassembling. Now, because of this reassembling, there are four non-identical haploid "grandchild" cells, each of which have less aged genetic material then their original progenitors. Parenthetically, the reassembling of genetic material is sufficient to classify this as a sort of "sexual" reproduction. See Ramesh, et al. (2005) for more on this topic.

[11]See David Lewis (1976, pp. 24–29), Derek Parfit (1984, pp. 261–266), David Wiggins (1979), Bernard Williams (1960/1973), and Robert Nozick (1981, part I, Chapter 1, pp. 27–114), who each discuss this topic, or matters closely related.

of your brain, leaving the other half for you. Changes in the three possibilities follow from this: (1a) You and your clone are alive, each with half of your brain; (2a) You and your non-carbon-based replica are both functioning, each with half of your brain; (3a) You and the formerly brain-dead other, your identical twin, are both alive each with half of your brain.

The next variation is a continuation of the situation just above. Your heart does give out, and so the half-brain that had remained with you is now to be transplanted. Here are the various possible surgeries: (1b) Your half-brain is transplanted into another clone of you, identical to the first one. (2b) It is transplanted into another non-carbon-based replica of you, also identical to the first non-carbon-based replica of you. (2c) It is discovered you were actually not a twin but a triplet, and the third triplet has suffered the same fate as the second one, namely a sudden brain death. Now your half-brain is transplanted into that third (formerly brain-dead) triplet, so now both of the two surviving triplets are alive and well, each with a fully functioning half of your brain.

Several pressing questions arise. Under which situations are you surviving as you, yourself? If your original body is dead, but your brain is functioning inside a clone, or a replica, are you surviving as you but in a new body? What about if your heart held out such that you were still alive with half of your brain, while the other half went to a clone, a replica, or a twin? Can they not be surviving as you, as long as you are still alive? But then what about when you die? Now are they you, when before they were not? Finally, the case most like that of amoeba fission: Your original body is gone, but two bodies, either clones, replicas, or two of your triplet sibship, each have half of your brain. Do you survive now as two of you? But then suppose then that one of the triplets dies shortly after the transplant? Does the fact that there is again only one, change the you-status of the remaining triplet? Or in retrospect, does it change the you-status of the two of the triples, when there were two?

For now, these questions will remain unanswered, including their relevance for the three foundational philosophical questions brought to light with the Trans experience. We will, however, return to consider these matters later, in Chapter 7. There, with a new focus and novel insights, several empirical Experimental Philosophy (X-Phi) experiments derived from classical Thought Experiments about the *Self* will be presented. These new experiments were adapted with the express purpose of addressing important implications for Trans persons. But before we get to the Chapter 7's Experimental Philosophy empirical experiments, let me present a number of additional classical Thought Experiments on the *Self* upon which the empirical versions are based.

2.2 Body/Brain Exchange: Is My *Self* My Psychology? Or, Is My *Self* My Brain?

Bernard Williams (1970/1973, p. 47) devises a Thought Experiment[12] in which two people—Person-A and Person-B—undergo what amounts to an exchange of brains; although from a slightly different perspective, one could equally frame the operations as a brain transplant and a total body transplant. Here is how Williams sets it up: Before the operation A and B are both told that after the body/brain exchange one of them will experience extreme pain, while the other will have no pain and in addition will receive a great deal of money. They are each asked which body and which brain they would prefer to endure the pain; and which they would choose to have the good fortune of no pain and more money. Suppose that Person-A opts for future pain in the A-Body which will then have the B-Brain; and Person-B also opts for future pain in the A-Body which will then have the B-Brain. Given these choices, the surgeon will be able to satisfy both patients with their pre-transplant wishes. And now, we can examine how each will fare.

Clearly, assuming that both Persons-A and -B would like the non-painful financial reward for themselves, we can see that the pre-exchange Person-A locates his/her "me"-ness in the brain, whereas the pre-exchange Person-B has his/her "me"-ness in the body. Which of these positions makes more sense? Which patient does better? Williams suggests that mostly it has been assumed that Person-A's view, that "me"-ness is in one's brain, is more rational (1970/1973, p. 51); and indeed, after the surgery the person with the B-Body, and the A-Brain, a lot of new money, and no

[12]Note that Williams attributes the origin of this Thought Experiment to Sidney Shoemaker (1963, p. 23.)

pain, would seem to agree. Even the post-surgical B-Brained/A-Bodied person, feeling pain and having no financial compensation, would have to admit to the mistake. Or so it seems!

Williams, however, gives a striking case example analysis that argues strongly for the contrary position (1970/1973, pp. 51–52). In this example he is able to demonstrate that considering the brain as the center of one's psychological "me"-ness, is, at the end of the day, not all that compelling. Put another way, Williams shows that there is a lot more contributing to one's "me"-ness than the contents of one's brain. I will abbreviate his argument, but continue to use most aspects of the original Thought Experiment involving Person-A and Person-B and their body/brain exchange, and stipulate, along with Williams, a few alterations.

Suppose now that I am Person-A. The surgeon tells me a few things about the operation: (1) The brain exchange will be that of brain software, rather than hardware, such that all of my brain-mediated content, mentation, and psychology will be transferred to Person-B's brain hardware, and vice versa; and (2) that sometime in the very near future, in preparation for the body/brain exchange procedures, my body will be in a lot of pain. It is true that at that time all of the contents of my brain will be on the way to being exchanged and transplanted into the hardware of Person-B's brain (and body), just as Person-B's brain contents will be on its way to my physical brain (and body). So, what this means is that at just about the time that the severe pain in my body will begin, my brain (the A-Brain) will lose all of my memories, and all of my software content. In fact, as the pain in the A-Body will continue and probably increase, all of the A-Brain's (my brain's) psychological connectedness to me-as-the A-Body will be halted. At this point, and until the A-Brain hardware receives Person-B's psychological software, it will in effect have no psychological contents at all. Should I (Person-A) feel heartened by this?

Williams (1970/1973) gives a straightforward answer. Due to what Person-A can anticipate experiencing, Person-A should certainly not feel heartened at all. Placing himself in Person-A's shoes and referring to himself as Person-A (as I did just above), Williams explains his view. Says Williams, it will be just like "…being involved in an accident…as a result of which I wake up in a completely amnesiac state [having no thought contents] and also in great pain" (p. 52). Next, according to the surgeon's description, the A-Brain (the brain Williams refers to as his own) will first be emptied, and then connected up with all of B's memories and B's other psychological content, as the pain in A's body will continue to intensify, reaching the level of torture. Should I (Person-A) before the exchange operation now feel relieved that all of the pain the A-Body (a.k.a. my own body now) will experience and endure, will be grappled with by a brain that will at that time be psychologically B's brain? Again Williams (1970/1973, p. 52) says emphatically no: "Fear, surely, would be the proper reaction and not because one did not know what was going to happen, but because in one vital respect…one did know what was going to happen—torture which one can indeed expect to happen to [one's body, i.e.] oneself …".

With this line of reasoning, based on the Thought Experiment, Williams (1956–1957/1973, p. 1) makes good on what had been his longstanding aim, namely to "…try to show that bodily identity is always a necessary condition of personal identity." This is an argument for physicalism about "me"-ness, extending beyond the status of the brain.

Thomas Nagel (1986, p. 42) endorses physicalism too, but takes the opposite tack, embracing specific brain-based physicalism. Arguing that although in the above Thought Experiments one may be able to imagine that one is still oneself, divorced from one's brain, if it is the case that one cannot really be oneself without one's brain, such imagining will show these imaginers to have "…confused epistemologic with metaphysical possibility." He continues, "In trying to conceive of my survival after the destruction of my brain, I will not succeed in referring to myself in such a situation if I am in fact my brain" (p. 42). For Nagel then, it is the case that he cannot be Thomas Nagel without his brain. When he cares about his survival, it cannot be without his brain (p. 42): "…if a physically distinct replica of me were produced who was psychologically continuous with me though my brain had been destroyed, it would not be me and its survival would not be as good (for me) as my survival."

Nagel appears to be directing these arguments to the position of Derek Parfit. Parfit holds that it is (or at least should be) a combination of both our psychological continuousness and our psychological connection (in

Parfit's terms, our "Relation-R") that matters when we are concerned about our survival. For Parfit it is Relation-R, more than personal identity, including body- and/or brain-based personal identity, about which one ought to care, when one is concerned about self-survival. I turn now to Parfit's interesting account, particularly a group of highly illustrative Thought Experiments he constructed in order to demonstrate his point. Again, I hold to the promise that that the implications for Trans persons[13]—regarding both the Parfit thought experiments to follow immediately, and the Body/Brain exchange experiments just discussed—will be taken up in detail in Chapter 7, the chapter on empirical Experimental Philosophy (X-Phi) studies adapted from these several classic Thought Experiments on the *Self.*

2.3 Tele-Transporting, Physical and Psychological Spectra: "Relation R"-*Self* vs. Personal Identity-*Self*

Derek Parfit (1984, pp. 199–201) uses a science fictional device, tele-transportation, in order to engagingly elaborate his views on (a) what constitutes the *Self* and personhood, and (b) what we do and should care about when we are interested in our own survival. In tele-transportation, a person, let's call him D, presses a button to signal his readiness to begin the process. Then, after some short time has elapsed, D is rendered unconscious. He next undergoes a procedure where his body, including his brain, is totally destroyed, but then body and brain are reconstituted on Mars. This takes place on a cell-by-cell basis, as the status of each cell is registered on Earth and then programmed into organic material available on Mars such that Mars-D is indistinguishable from Earth-D both physically and psychologically. (Parfit is operating with a physicalist brain-based psychology here.) The Earth version of Person-D never

[13]Here is admission I offer in advance: While the Experimental Philosophy studies will offer clear implications for Trans persons, and will have other demonstrable advantages, answers to the three foundational and vexing questions about the *Self* raised by the Trans experience will still remain largely out of reach.

regains consciousness, while the Mars version experiences himself as D, physically and psychologically, with no doubts at all.

On Parfit's view, Person-D should not have much trouble with this. Mars-D is a perfect physical replica of himself, and no less organic; moreover Mars-D is totally psychologically connected and seamlessly psychologically continuous with Earth-D. And to the extent that Person-D embraces either a Personal Identity view of *Self* or Parfit's position—namely that it is Relation-R, consisting of both psychological connection and continuity of his body and his brain's contents, that contribute what is essential to survival—both the Earth-D and Mars-D instantiations of Person-D, should have no trouble whatsoever with tele-transportation and preserved *Self* identity. In fact, Parfit "reports" that Person-D makes several round tele-transporting trips, without incident or consternation.

Since these science fiction Thought Experiments are designed to elaborate Parfit's account of Relation-R, which he wants to contrast with the Personal Identity-*Self* view, let us outline them both now, before any more journeys. Relation-R for Parfit (1984, p. 206) consists of psychological connectedness—"the holding of particular direct psychological connections," and psychological continuity—"the holding of overlapping chains of strong connectedness." Psychological connectedness can be seen in overlapping chains of experience: in memories, in continued beliefs and desires, in enduring personality traits, and particularly in the carrying out of intentions (pp. 205, 207). Personal Identity, as it is usually described also comprises one's memories, but also current sensations, perceptions, emotions, and includes input from many sources: one's body, one's self assessments, other people's observations, one's ideas, one's history—basically much of the software content of *Self*. Parfit (1984) distinguishes Personal Identity from Relation-R as follows. Relation-R features strong connectedness as an essential central aspect; and neither Relation-R nor strong connectedness is transitive. Thus, X is strongly connected to him/herself yesterday, but not to him/herself forty years ago; and X ten years hence is not as strongly connected to X now, as X now is to the X of tomorrow. Personal Identity, on the other hand, is transitive. X is X now and was X forty years ago and will be X in ten years. Continuing to explicate the distinction, Parfit states: "Because identity is a transitive relation, the criterion of identity must also be a transitive relation. Since strong connectedness [Relation-R] is not transitive, it cannot be the criterion of identity" (p. 206). In other words, Relation-R is not only different from Personal Identity, it cannot be the criterion by which identity is determined. And, as will be described later below, Parfit maintains that it is Relation-R, rather than Personal Identity, with which he is concerned when he is concerned with his own survival. Moreover, and not surprisingly, for Parfit, Relation-R is not only different from Personal Identity, it is more important.

The difference between Relation-R and Personal Identity is generally unproblematic for most people—including those who are tele-transported, like Person-D—because there is usually considerable overlap between Relation-R and Personal Identity. However, there are situations where Relation-R and Personal Identity are not coincident. It is these cases that provide some interesting tests for the two distinct views of the *Self*, when it is self-survival that is at stake. Here is one such situation, the "Branch-Line case" as Parfit (1984, pp. 200–201) imagines it: Tele-transportation has been improved such that Earth-D will no longer be destroyed. Person-D is informed of this on his latest trip, but only after he's already pressed the button thereby initiating his tele-transportation. D immediately realizes that his earthly and mars-ly counterparts will soon co-exist and even communicate, suddenly raising a big problem. He, now (on earth) feels his Personal Identity is as Earth-D; but Mars-D will also identify himself as D. Moreover, Mars-D will be fully psychologically connected, and absolutely continuous with Earth-D, to all appearance identical. But can there be two of him? Now as if that didn't complicate things enough, D is next told that the new technology has failed in a peculiar way; D is made aware that although indeed his body (a.k.a. Earth-D) won't be rendered unconscious and immediately destroyed in the usual fashion, he will suffer a slow cardiac death in a few days. Will this change how Earth-D feels about Mars-D as a continuation of himself? In any case, in a few days there will no longer be two Ds, but the one remaining will be Mars-D.

With the old tele-transport system, D's life could plausibly be construed as progressing along one main line; Person-D went from Earth-D to Mars-D. But with the advent of the so-called improved technology,

D's life could continue along two branches simultaneously. Both Earth-D and Mars-D would exist; and they would survive in parallel. True with the spectacular technical failure of the new procedure, only Mars-D will continue to exist, as Earth-D will die; but this will follow a possible troubling, albeit short, period of parallel existence.

These imagined events raise several philosophical problems: First, for those who initially regarded Mars-D as a continuation of Person-D, can they still regard him as such when there is a parallel existence with Earth-D? If they can, they are left with the problem of D being more than one person. And how can something that is numerically identical—Person-D is identical to himself—be two entities? If this problem abates for them when Earth-D dies—for then Mars-D will be the sole Person-D—they must account for a serious inconsistency: how can contingent matters affecting Earth-D alone entirely change the status of Mars-D as identical with Person-D?

Obviously, these Branch-Line issues are much like those discussed above in the subsections on fission. They are well recognized by various philosophers discussed (or to be discussed) in this section: Parfit (1984, pp. 261–266), Wiggins (1979), Williams (1960/1973), and Nozick (1981)—the last of whom (p. 32) sums up the situation in this way: "…identity cannot depend upon whether there is or isn't another thing of a certain sort; …if there could be another thing so then there would not be identity, then there isn't identity, even if that other thing does not actually exist."

But according to Parfit (1984, pp. 202–209, 231–243) things are no simpler for those who never regarded Mars-D as a continuation of Earth-D. Noting that a butterfly developing into a caterpillar is a naturally occurring example of a numerically identical creature, yet one that is very different in each stage, Parfit considers questions about the limits of what can constitute a person's continuance as that same person. Suppose that cells from Earth-D were replaced with the material from which Mars-D's brain was to be constructed, not only one neuron at a time, but very slowly. Would the gradual versus the sudden replacement cause those who viewed Earth-D and Mars-D as merely "exactly similar" to change their opinion so that Earth-D and Mars-D would now be deemed "numerically identical," with Mars-D as the rightful singular continuer of Earth-D as Person-D? Or a simpler question, how many replacement parts can something endure and still be the original? Personhood, Parfit concludes, seems to be determined by (1) physical criteria, amounting to spatio-temporal physical continuity of at least the brain and at most the brain and body, and/or (2) psychological criteria, namely the psychological continuity and psychological connectedness of Relation-R. (Remember, that for Parfit, Relation-R, although psychological in nature, is itself physically based, specifically brain based.)[14]

But then Parfit takes a very interesting turn. Using the notion of spectra, physical, psychological, and physical/psychological combined, Parfit demonstrates with sorites arguments[15] that all of the concepts involved in evaluating personhood are hopelessly vague. Would you accept that someone is still you, if 10% of your brain had been removed, and you were left with 90% of your original physical make-up, and 90% of your psychology? You can also ask the question this way: Is a person who is 90% you, both physically and psychologically, you? What about 88% or 85%? And then there is this further question: Does the person who is 85% of your physical being and 85% of your psychology still the "you" you want to survive when you care about your survival? What about 80%, or 75%? If you accept 75%, what about 68%? And so forth. What Parfit shows is that there is no clear, coherent concept of person, using physical, psychological, or combined measures.

We are left with two linked problems: (1) the fission problem—when what I care about is my own survival, how can there be two of "me"? And (2) the vagueness problem—how can I demarcate what constitutes "me", when the concept is vague, physically and psychologically? Parfit (1984) has an original solution that works for both problems. He suggests (pp. 233, 235, 241–243), that although we want to think of our identity as determinate, and we want to be able to answer questions like: "Am I going to die in the tele-transporter?" and "Is that person me or not me?"—these questions are empty and wrong. To remedy this, he

[14]On Parfit's account whatever is important in one's psychology—one's intentions, desires, goals, personality, etc.—all of this is caused by brain goings-on.

[15]For an introduction to sorites arguments, and to vagueness, See Brakel (2010, Chapter 3, pp. 53–87) "The limits of rationality: vagueness a case study." This chapter also includes many basic vagueness references for those interested.

suggests that when I am interested in my survival, I should be less concerned with perpetuating my particular singular Personal Identity and more concerned about the survival of all and anybody (anything) with Relation-R to me. As long as my psychological continuity and connectedness continues, it doesn't matter if I have undergone fission and there are two (or even more) of me. Relatedly, I do not have to specify with precision what (who) would count as someone with my Personal Identity, i.e. the "same person as I am." Instead, given that I am concerned with my own survival, I should be increasingly concerned with the survival of those who have more Relation-R to me; and I should still have some interest in anyone who has some (or any) Relation-R to me, etc. In other words, for Parfit the matter is relatively straightforward and not all that complicated: My concern with my own survival should essentially be a concern with the survival of all Xs that bear Relation-R to me; and my concern with any particular X should vary positively as a function of that X's increased Relation-R to me.

Note that in Parfit's solution to the dilemmas of the survival of "me"-ness, Personal Identity is supplanted by Relation-R. This is interesting, and so is one of its consequence—that singularity no longer has overriding importance. Nozick, whose views we will discuss next, also treats the singularity aspect of "me"-ness in an unusual fashion. But, before we leave Parfit, we should note that the Parfit's Relation-R view does provide another potential avenue of investigation regarding the *Self* of Trans persons: Namely, in order for Trans person to feel less "wrong," perhaps more self-connected if not self-identical, how much gender change would be wanted?[16] We will not explore matters such as these here. Instead, they will be taken up at length in Chapter 7, which consists of several empirical Experimental Philosophy Thought Experiments derived from the classics we have been discussing.

---

[16]Parfit's Relation-R, particularly the psychological connectedness aspect, might add to our understanding of evolving *Self* feelings and perhaps even increasing *Self* satisfaction when Trans persons transition. Thus, returning to our emblematic GB, as a transition progresses GB might feel greater psychological connectedness with some basic GB-*Self*. The psychological continuity part of Relation R, on the other hand, would not seem to similarly benefit.

Note too however, that Parfit's Relation-R, as for all the accounts explored so far, cannot supply solutions to the puzzling ontological and epistemological questions about *Self* occasioned by the trans phenomenon.

2.4 Closest Continuer

For Nozick (1981, pp. 33–70) it is our closest continuer about which we are concerned, when we are interested in our own survival. However, if we have no closest continuer, but instead have more-than-one close enough continuers, we can care about these just about as much. Defining the closest continuer, Nozick (1981, p. 34) explains: "The closest continuer view presents a necessary condition for identity; something at t_2 is not the same entity as x at t_1 if it is not x's closest continuer. And 'closest' means closer than all the others. If two things tie [in their closeness to x]…then neither is the same entity as x." Nozick next points out that while closest continuer status is necessary for identity constitution, it is not sufficient: "…something may be the closest continuer of x without it being close enough to be x" (p. 34).

Nozick goes on to describe several interesting thought experiment cases, variations on those we have explored, involving transplanted brains and hemi-brains, cloned bodies, dying original bodies, and replicas with psychological continuity. What is interesting in Nozick's examples is that depending on the circumstance, the very same person (or entity) that is considered the closest continuer in one situation will certainly not be the closest continuer in another. To illustrate, take Parfit's tele-transportation type cases. As long as Earth-D is alive, Nozick asserts that Earth-D and not Mars-D, will be Person-D's closest continuer. However, as soon as Earth-D is gone, either through immediate destruction in the original process of tele-transportation, or via the slow, cardiac death days later caused by the new procedure, Mars-D becomes Person-D's sole closest continuer. Nozick has in this way, with the context sensitivity of the closest continuer model, been able to solve elegantly several of the vexing dilemmas about "my" survival that we've been considering. Earth-D is the defacto Person-D until Earth-D is no more. Then Mars-D becomes Person-D, despite the fact that this was not Mar-D's status just moments prior to Earth-D's destruction or death. When tele-transporting Person-D back to Earth, Mars-D ceases to be Person-D as soon as Earth-D is back on the scene. If something goes awry and Earth-D is never reconstituted as such, Mars-D remains as Person-D. Indeed, Nozick's solution works very well, particularly in addressing questions like the one articulated above: How can contingent matters affecting Earth-D alone entirely change the identity status of Mars-D with respect to Person-D? Dilemmas like this dissolve with Nozick's context dependent closest continuer. However, the closest continuer model does face a serious problem when it comes to those cases in which there are ties as to which of two (or more) possibilities is someone's closest continuer.

"Y at t_2 and Z at t_2 continue X at t_1 equally closely, closely enough so that either, in the absence of the other, would be (part of the same continuing entity as) X at t_1" (Nozick 1981, p. 62). Considering himself X and, no longer existing, given that there is no single closest continuer, Nozick (p. 63) writes, "I am neither Y nor Z, and [therefore] I no longer exist. This is not distressing in this case, for what I care about is that there remains something that continues me closely enough to be me if it were my sole continuer; and if there are two such, I care especially about the fate of the [tied] closest continuer." But this suggests that the closest continuer can be two. But Nozick holds that neither identity over time nor personal identity can be two (p. 68), and he has further stated (p. 68): that "the closest continuer better realizes identity than does the relation: tie with close enough continuers." To further complicate matters, Nozick remarks several times (pp. 64–65, 67, 68) that although we care most about the closest continuer when there is one, we do not care all that much that there is one, as long as there are continuers that are close enough: "[G]iven that we care about our identity, we care especially about our closest continuer when it exists, yet do not care especially that we have a closest continuer, provided we have close enough continuers. How is the tie case to be described on this view? I do not view a tie as like death; I am no longer there, yet it is a good enough realization of identity to capture my care which attaches to identity" (Nozick, 1981, p. 68).

What are we to conclude? Either Nozick is claiming that identity is not numerically singular despite his own comments to the contrary and the threat of incoherence; or he is claiming there is something else—an identity-like relation, consisting in either a closest continuer or close enough continuers. Neither of the latter two requires singularity, rather a good enough realization of my identity to sufficiently constitute my "me"-ness when I am concerned with my own future survival.

Again, we shall not take up Trans implications of the Nozick view until Chapter 7 on empirical Experimental Philosophy Thought Experiments. But meanwhile, there is something unsatisfying with this conclusion. For one thing it leaves open the possibility of "type" and "token" people. Described by Williams (1970/1973, pp. 79–81), a particular type-person could have any number of tokens, all of whom could count as close enough continuers. Here is how this would look. Take the brain of someone, me, LAWB. Now take all of its structure and information and transfer it cell-by-cell to a new brain; and put this new brain in a new body built up also on the blueprint of LAWB. But if this can be done once it can be done n-number of times. This seems like a reductio argument against Nozick's position, as Nozick would have to claim that as the original LAWB, I, should be fine with this whole bevy of "me"-tokens, and not especially prefer that I should continue as just one. Even if this were unproblematic, one would still have to wonder in cases like this with multiple and tied closest continuers, what exactly Nozick has in mind that would (a) be distinct from personal identity and identity over time and yet (b) be capable of instantiating identity well enough to (in his words, p. 68) "capture my care". What is this identity-like concept?

Rather than trying to resolve Nozick's problem in reconciling identity or "me"-ness with his version of non-singularity, let me suggest that the fact of his serious troubles can prove highly instructive. In fact, next I will take up "me"-ness as necessarily singular, the efforts of Parfit and Nozick to the contrary notwithstanding. In order to make this case I will continue with various philosophical dilemmas and accounts.

2.5 My "Me"-Ness, My *Self* Is My Agency

When I am interested my own survival, what is the nature of this "me" with which I am concerned? I propose the following answer: The "me" whose survival interests me when I am concerned about my own survival is the singular "me" who represents my agency. This answer leads directly to the next question—just what is agency, what is an agent? Most notable in any understanding of the characteristics and capacities of agents are their intentions and intentional acts; these predicated on their desires, beliefs, other psychological attitudes, as well as personality and character traits. Among the many desires I have, some are universal (or nearly so) as they are related to the drives striving to satisfy biological needs. When I am thirsty, I desire to drink; when hungry, I desire to eat, etc. Also, along with many (most) other animals I have the drive-derived desire to continue living,[17] this desire (or actions that imply this desire) being the very one that fuels my concern about my own survival. Note that these drive-derived desires in particular demonstrate clearly that my agency is different from my personhood. The latter is more sophisticated, species-specific with respect to highly developed moral and rational concerns, and thereby distinct from other non-human animals. My agency, on the other hand, predicated on drive-derived desires common throughout the animal world,[18] is closely tied to action. The roles of action and intention in agency, and a continued discussion of singularity, follow immediately below.

If I am correct that it is oneself as a singular agent about whom one is concerned when one is interested in one's own survival, we should see a problem with contemplating ourselves as more than one living entity. Thomas Nagel (1971) highlights the problem. Describing commisurectomy patients—those who have had one half of their brain removed owing to tumor or seizure—Nagel admits that many seem to function in a variety of circumstances as though with two separate consciousnesses. And yet Nagel cannot find agency anything other than singular, even in these special cases, and therefore remarks, confused and confusingly: "[W]e take ourselves as paradigms of psychological unity, and are then unable to project ourselves into their lives, either once or twice" (Nagel, 1971, p. 410). Later, Nagel (1986) better explicates his singular identity view: "If I am told that my brain is about to be split, and that the left half will be miserable and the right half euphoric, there is no form that my subjective expectations can take, because my idea of myself doesn't allow for divisibility—nor do the emotions of expectation, fear and hope" (pp. 44–45). For Nagel his agency, the "me"-ness from

[17]From a strictly biological point of view, one would have to say that the drive to continue living is shaped very much by the drive to live long enough to reproduce as successfully as possible.

[18]See Brakel (2016, pp. 1–3) for a view of animals as agents.

whence his subjective expectations, fears, and hopes arise, is necessarily singular.

David Wiggins (1979) also stresses singularity in agency, linking it to concerns about survival of the *Self*: "The instinct for survival has played its part…But what is the content of said instinct? The content is surely that this animal that is [numerically] *identical with me* should not cease to be but should survive and flourish" (p. 420) [Wiggins' italics]. Discussing later the distinctly human sorts of sublimated traces one can leave, i.e., living on in one's works, or having done some lasting good, Wiggins (1979, p. 422) could not be clearer. It is his singular agential *Self* whom he cares about in terms of his survival: "What I am certain about is that I do not see how the offer of any of these things, Parfitian [Relation-R descendants] or etiolated, can be taken for a proper surrogate (equivalent on the level of imagination, conception, and desire) for the continued existence of the one and only…me."

To explore the central role of intention in agency, we turn now to G. E. M. Anscombe's 1957 classic monograph, Intention. She asks (p. 9), "What distinguishes actions which are intentional [agential] from those which are not?" She responds to her own question: "…they are the actions to which a certain sense of the question 'Why?' is given application; the sense is…the answer…gives a reason for [the] action." Anscombe moreover considers this a necessary feature of [agent driven] intentional actions, "…the concept of …intentional action would not exist, if the question 'Why?', with answers that give reasons for acting, did not" (p. 34).

These definitions of agency seem fine for language competent humans. But I have held that the drive-mediated desires that we share with animals are quite essential in our agency. Indeed, in my view (as discussed earlier and argued elsewhere[19]) animals are agents too. Anscombe (pp. 86–87) also addresses this matter, concluding that the descriptions of many animal actions fit too well with descriptions of intentions, not to be intentions. Anscombe leaves it there; but owing to the work in evolutionary biology during the intervening decades, we can now also avail

[19]See Contrast Case 4—Non-human Animals in Chapter 2. For references to these earlier Brakel works see footnotes 9 and 18 in this chapter.

ourselves of a more complete understanding of agency, thanks to some basic evolutionary biology concepts.

To the extent that animals can have transparent motives, intentions, and intentional actions of the sort we humans have, the question "Why?" can be asked just as it can in the human case. But what about actions that are instinctually driven, but appearing to be irrational, such that the standard "Why?" question would seem to have no useful application? Take for example frogs indiscriminately swallowing bugs and black metal BBs. A case like this reveals that for non-human animals (and for humans too) motives, intentions, and actions often have both proximate and ultimate explanations (see Alcock 2001, pp. 16, 130).[20] The proximate explanations are those understood usually in terms of physiologic/functional goals, and can be seen as analogous to volitionally predicated answers to the "Why?" question. So, why does this frog swallow bugs and BBs? Proximally, because this frog is hungry. Similarly, why does the chicken cross the road? The proximal explanation: She wants to get to the other side. But, in the ultimate explanations, answers to the "Why?" questions are determined by evolutionary (i.e., selective reproductive) fitness success aims, rather than strictly volitional ones. Thus, there is an ultimate explanation—in addition to the proximal one—for why hungry frogs swallow all those metal BBs. The ultimate explanation: Frog reproductive fitness is improved by better nutrition, which in turn has been enhanced by a swallowing mechanism that is triggered whenever an object of a certain description—one clearly fitting both bugs and BBs—is but a tongue's length away. Rounding out the case for actions admitting of both proximate and ultimate explanations, let's return to the chicken. Yes, proximally she crosses the road because she wants to get to the other side; but ultimately, she might especially want to owing to the presence of food, or a potential mate, or because she is being chased by a predator, etc.—actions here all enhancing selective reproductive fitness.

With the question "Why?" extended such that it can include evolutionarily explanations as well as volitional reasons, agency can be extended throughout the animal world, whenever there are actions to

[20]Note, also that the original concept of these different sorts of explanations owes to Tinbergen (1963).

which one or the other application of "Why?" can pertain. Thus, on my account, even amoebas should be viewed as agents, as for example, they move to engulf or extrude. Why does amoeba X extrude that y-type particle now? On the evolutionary application of why: Amoeba X extrudes this particle now, because extruding particles of y-type has increased the reproductive fitness of amoebas who do so.[21] Also, why does amoeba X divide now into daughters A and B? Again, the evolutionary/ultimate explanation provides an answer. The selective fitness success of X's genetic material is enhanced, if division, upon which replication immediately follows, occurs at a particular time and when certain conditions obtain. In my view then, not only is it the case that amoebas are singular agents, but paradoxically they are single agents whose last agential acts under normal conditions are those involving their own division into two new agents.

We have talked of agency as singular, and agential actions as intentional in that they are actions to which the question "Why?," in at least one of its forms, can be applied and answered. And so now let's explore the movement aspect of agential actions. Agency implies motivated motion, i.e., intentional action. Linking singularity, action, motion, agency, and what amounts to the concern about my continued "me"-ness, Stuart Hampshire (1959, pp. 59–60, 67, 68–69) describes the initial conditions for being a person as entailing being a distinct physical entity who acts with movement, agentially on other objects and persons in the world.[22] Mirroring Anscombe, Hampshire (1959) claims, "'With a view to', or 'in order to', are unavoidable idioms in giving the sense of the notion of an action, the arrow of agency passing through the present and pointing forward in time" (p. 73). Moreover, for Hampshire (1959) the body-in-motion is not just an instrument of intentional acts (pp. 79–80); rather the body is the intention-in-action, such that when an action is animated by an agent's intentions, "there is no reason to look for some

[21]The explanation goes beyond this too: The selective fitness of amoebas inheriting the capacity to extrude y-type particles is also improved.

[22]Interestingly, for Hampshire, knowing oneself as a person, distinct from other persons and objects, entails noting the effects of one's own agential actions. This seems a very Kantian idea from the *Critique of Pure Reason* (1781/1787), where the unity of the *Self* rests upon the unity of objects and the perception thereof; while the unity of an object is known only through the several percepts of it all possessed by the same unified perceiving diachronic self.

criterion of personal identity that is distinct from our bodies as persisting physical objects" (pp. 74–75). In other words, our personal identity is no more and no less than our agential intentions, these intentions necessarily instantiated in our physical bodies. Finally, putting agency at the center of psychological continuity too, Hampshire (1959, p. 72) states, "…we carry our intentions with us, and this carrying forward of intentions, together with the perception of movement, provides the natural and necessary continuity of our experience."

Writing at about the same time as Hampshire (but with very different philosophical goals), P. F. Strawson (1959) comes to similar conclusions regarding action and agency. Strawson's work is a quest for a descriptive metaphysics of "individuals," both material entities and persons. Dividing up the world into the predicates pertaining to persons, P-predicates, and those pertaining to the material realm, M-Predicates, Strawson (1959, p. 104) discusses the problem of how P-predicates are even possible: "…a beginning can be made by moving a certain class of P-predicates to a central position…They are predicates…which involve doing something, which clearly imply intention…and which indicate a characteristic pattern…of bodily movement" (p. 111). (I take Strawson to be talking of agents and agency here, despite his use of the term "person" as he is at this point not requiring anything of his P-predicates beyond simple intentional activity.)

From yet another totally different philosophical quarter, Maurice Merleau-Ponty (1962, p. 137) in his *Phenomenology of Perception* presents a similar slant on agency, desire, and action. In a section in which he considers the role of our own movements in the development of concepts of space, he asserts, "Consciousness is in the first place not a matter of 'I think' but of 'I can'." This phrase is commented upon by A. J. Ayer (1982) who agrees with Merleau-Ponty and adds, "…this implies…that we enter the world as [active] agents as well as observers, and that what we attend to may at the most primitive level be a function of what we desire…" (pp. 218–219).

Finally, the biological foundations of agential action warrant our attention. Howard Shevrin (2003) has developed a view of agency not far from my own. He locates a particular part of a drive, its "pressure" or "motor factor"—that which Freud (1915, p. 122) characterized as "the amount of force or the measure of the demand for work which it [the drive] represents"—as the central factor both in action and agency. Shevrin (2003, p. 13) proposes, "that this pressure...is at the heart of what we mean by agency, the foundation for the...self as *actor* [Shevrin's italics]." Then Shevrin (p. 13) further suggests that "From a neurophysiological and neuroanatomical standpoint agency appears to be instantiated in a particular brain circuitry characterized by a particular neurotransmitter." Shevrin refers here to the "SEEKING system" described by JaakPanksepp (1998, Chapter 8, pp. 144–163), consisting of dopamine neurotransmitter (DA) circuits in the lateral hypothalamic corridor from the ventral tegmental area (VTA) to the nucleus accumbens septi (NAS) of the brain (Panksepp 1998, p. 145). Immediately upon electrical stimulation of this area, Panksepp (1998, p. 145) describes the reaction in animals as "...the most energized exploratory and search behaviors an animal is capable of exhibiting." As this SEEKING system activity is not linked to a specific drive, and because (chemically) blocking the DA NAS/SEEKING system circuitry causes inertia even under conditions of need, Shevrin (2003, p. 13) suggests that its activation is "a pure culture of agency", that the "urge to move, to act is the most primitive and abiding aspect of animal life..." and that "Self-initiated movement and action *is* agency [Shevrin's italics]." Now, while I take issue with him here—for on my view agents must not only act, but act in such a way that one or another version of the question "Why?" must be applicable—we are certainly in agreement that all animals are capable of agent-driven actions. And, Shevrin's insight has provided reason to believe that the motor aspect of agential action has a specific brain based biological foundation.

Further biological matters will form the core of Chapters 5 and 6. But, so as not to get out too far ahead of ourselves, there is much that remains before leaving this chapter; in particular more on the Agential View of *Self*. Just below, and relevant to questions of agency, I present Contrast Case 7—Amnesia. (After that, subsections on "Problems with the Agential views of *Self*," and "Implications of the Agential views for Trans persons and the Foundational Questions," will follow, respectively.)

Contrast Case 7—Amnesia

Amnesia can be conceptualized as a central loss of memory, but along with this, given the importance of memories in many accounts of the *Self*, some degree of damage to one's Personal Identity-*Self* might be expected. Undeniably, amnesiacs lose track of who they are writ large and small. They can forget their own character traits and preferences, what they had deemed important, what they care about, and even whom they love and like and why. There are reports (perhaps merely anecdotal) that dementia patients "forget" their gender. Along these lines, but fully referenced, here is the unusual story of ABC News producer, Don (then Dawn, but later Don again) Ennis. Reported in a *New York Magazine* article (byline Joe Coscarelli, August 6, 2013) Ennis, who had transitioned to female, suffered a bout of transient global amnesia, during which time "she" not only forgot, but then also denied ever having transitioned to Trans-female status, resuming "his" maleness thereafter.[23]

Consider now someone who has lost all memory, and thus likely his/her identity. When this person, X, states: "I am now moving my arm." or "I am in pain," X is still a viable agential-*Self*, making first person statements immune from error through misidentification (Wittgenstein 1958; Shoemaker 1968).[24] Similarly, Anscombe (1974/1994, p. 155) describes the Total Amnesiac, who, while retaining no memories, and nothing in the way of personal identity, is nonetheless intact in his bodily agential actions. Consequently, first person statements such as "Yes, I will walk over there" and "I'm getting up out of the chair now" are true, and as immune to error through misidentification as would be the case with persons of normal memory and normal personal identity awareness. Anscombe has thereby demonstrated the self as agent: Like X, Anscombe's Total Amnesiac has lost all memory and identity, but not his agential-*Self*, his agential "I".

---

[23]Ennis's case does suggest that some intact memory capacity is necessary for maintaining gender identity—both of the Cis and Trans type!

[24]The phenomenon is well described by its name, but see Wittgenstein (1958) and Shoemaker (1968), for more discussion on the concept "immune from error through misidentification."

Problems with the Agential-*Self* View

Significant in itself as a Contrast Case, the phenomenon of amnesia reveals a serious problem manifest in the Agency-*Self* view. I've presented the circumstance, just above, in which a person, one even suffering from total global amnesia, can continue to function as a viable Agential-*Self*, answering certain "Why" questions of both the evolutionary and reasons-driven sort. But these reasons must be thin and narrow indeed, especially in light of the amnesiac having lost important aspects of the Personal Identity-*Self*—namely, his/her central cares, goals, desires, and especially memories. That all of this can be so compromised can only mean that the Agential-*Self* is rather bare, explaining only so much.

Implications of Agential-*Self* View for Trans Persons and the Foundational Questions

The Agential-*Self* view does provide insight into the nature of the "me" I'm concerned with when I'm concerned with my survival. But as important as this is, and as broad in terms of applications for non-human animals, regarding the fundamental philosophical *Self* questions the Trans phenomenon raises, its great breadth should be contrasted with its extreme narrowness. Incontrovertibly, the Agential-*Self* view makes clear that Trans persons are singular agents before, during, and after transitioning. And, this is true even in the extraordinary case presented just above where the person in question first forgot, then denied, then no longer wanted the transition. As with all agents, many (most) of the acts of Trans agents are immune from error through misidentification—the agential "I," the agential-*Self*—ever active. But beyond this, the Agential-*Self* view provides no further understanding of what aspect(s) of *Self* are at issue, either epistemologically or ontologically, operating in the fully agential people who want to, and then do undergo gender transition.

There is however something to add. As was speculated regarding Buss's "normal human flourishing view of self-constitution" (see the earlier section in this chapter), the Agential-*Self* account may also point the way to a potential biological proper-function account relevant for Trans persons. If so the Agential-*Self* view may yet yield some answers to some less bare, non-trivial "Why" questions of the evolutionary sort. Just such a proper-function account for gender transition, will be offered in Chapter 5.

Bibliography

Alcock, S. (2001). *The triumph of sociobiology*. Oxford: Oxford University Press.

Anscombe, G. E. M. (1957). *Intention*. Cambridge, MA: Harvard University Press.

Anscombe, G. E. M. (1974/1994). The first person. In Q. Cassam (Ed.), *Self knowledge* (Chapter 8, pp. 140–159). Oxford: Oxford University Press.

Arpaly, N. (2003). *Unprincipled virtue: An inquiry into moral agency*. Oxford: Oxford University Press.

Arpaly, N., & Schroeder, T. (1999). Praise, blame and the whole self. *Philosophical Studies, 93,* 161–188.

Ayer, A. J. (1982). *Philosophy in the twentieth century*. New York: Vintage Books.

Brakel, L. A. W. (2010). *Unconscious knowing and other essays in psycho-philosophical analysis*. Oxford: Oxford University Press.

Brakel, L. A. W. (2016). Animals are agents. Commentary on Rowlands: Animal personhood. *Animal Sentience, 103,* 1–3.

Brakel, L. A. W., & Fileva, I. (unpublished ms). *A new view of self-constitution*.

Buss, S. (2012). Autonomous action: Self determination in the passive mode. *Ethics, 122,* 647–691.

Butler, J. (1726). *Sermon II: Upon the natural supremacy of conscience*. Fifteen Sermons. Cambridge: Hilliard and Brown.

Chen, S., Urminisky, O., & Bartels, D. (2016). Beliefs about the causal structure of the self-concept determine which changes in personal identity. *Psychological Science, 27,* 1398–1406.

Frankfurt, H. (1971/1988). *The importance of what we care about*. New York, NY: Cambridge University Press.

Frankfurt, H. (1992). The faintest passion. *Proceedings and Addresses of the American Philosophical Association, 66,* 5–16.

Freud, S. (1915). *Instincts and their vicissitudes* (The Standard Ed., Vol. 14, pp. 111–140, J. Strachey, Ed. & Trans.). London: Hogarth Press, 1957.

Hampshire, S. (1959). *Thought and action.* New York: Viking Press.

Kant, I. (1781A/1787B/1965). *Critique of pure reason* (N. K. Smith, Ed.). New York: St. Martin's Press.

Kant, I. (1785/1997). *Critique of practical reason* (M. Gregor, Trans. & Ed.). Cambridge: Cambridge University Press.

Kant, I. (1785/1998). *Groundwork of the metaphysics of morals* (M. Gregor, Trans. & Ed.). Cambridge: Cambridge University Press.

Kant, I. (1793/1960). *Religion within the limits of reason alone* (T. Greene & H. Hudson, Ed. & Trans.). New York: Harpertorch Books.

Korsgaard, C. (1996). *The sources of normativity.* Cambridge: Cambridge University Press.

Korsgaard, C. (1999). Self-constitution in the ethics of Plato and Kant. *Journal of Ethics, 3,* 1–29.

Lewis, D. (1976). Survival and identity. In A. Rorty (Ed.), *The identity of persons* (Chapter 1, pp. 1–40). Berkeley and Los Angeles: University of California Press.

Merleau-Ponty, M. (1962). *Phenomenology of perception* (C. Smith, Trans.). London: Routledge.

Nagel, T. (1971). Brain bisection and the unity of consciousness. *Synthese, 22,* 396–413.

Nagel, T. (1986). *The view from nowhere.* Oxford: Oxford University Press.

Nozick, R. (1981). *Philosophical explanations.* Cambridge, MA: Harvard University Press.

Panksepp, J. (1998). The SEEKING system and anticipatory states of the nervous system. In *Affective neuroscience: The foundation of human and animal emotions* (Chapter 8, pp. 144–163). New York: Oxford University Press.

Parfit, D. (1984). *Reasons and persons.* Oxford: Clarendon Press.

Penelhum, T. (1971). The importance of self-identity. *Journal of Philosophy, 68,* 667–678.

Plato, A. (380 B.C. approx./1997). The Republic (G. Grube, Trans.). In J. Cooper (Ed.), *Complete works of Plato.* Indianapolis: Hackett Publishing Company.

Ramesh, M., Malik, S., & Logsdon, J. (2005). A phylogenomic inventory of meiotic genes; evidence for sex in Giardia and an early eukaryotic origin of meiosis. *Current Biology, 15,* 185–191.

Scanlon, T. (1999). *What we owe each other.* Cambridge, MA: Harvard University Press.

Shapiro, T. (1999). What is a child. *Ethics, 109,* 715–738.

Shevrin, H. (2003, September 15). *The psychoanalytic theory of drive in light of recent neuroscience findings and theories.* Wilson Memorial Lecture, New York University, New York.

Shoemaker, D. (2003). Caring, identification, and agency. *Ethics, 114,* 88–118.

Shoemaker, S. (1963). *Self-knowledge and self-identity.* Ithaca and New York: Cornell University Press.

Shoemaker, S. (1968). Self-reference and self-awareness. *The Journal of Philosophy, 65,* 555–567.

Sripida, C. (2016). Self-expression: A deep self theory of moral responsibility. *Philosophical Studies, 173,* 1203–1232.

Strawson, P. F. (1959). *Individuals.* London and New York: Routledge.

Strohminger, N. (2017). The true self. *Perspectives on Psychological Sciences, 12,* 551–560.

Tinbergen, N. (1963). On aims and methods of ethology. *Zeitschrift für Tierpsychologie, 20,* 410–433.

Watson, G. (1975). Free agency. *The Journal of Philosophy, 72,* 205–220.

Wiggins, D. (1979). The concern to survive. *Midwest Studies in Philosophy, 4,* 417–422.

Williams, B. (1956–1957/1973). Personal identity and individuation. In *Problems of the self* (Chapter 1, pp. 1–18). Cambridge: Cambridge University Press.

Williams, B. (1960/1973). Bodily continuity and personal identity. In *Problems of the self* (Chapter 2, pp. 19–25). Cambridge: Cambridge University Press.

Williams, B. (1970/1973). The self and the future. In *Problems of the self* (Chapter 4, pp. 46–63). Cambridge: Cambridge University Press.

Wittgenstein, L. (1958). *The blue and the brown books.* Oxford: Blackwell Press.

4

Selected Contemporary Gender-Theorists' Views; or Where Is the *Self*?

Abstract Surprisingly, questions about the concept of *Self* as framed in the monograph, are not uppermost in this important and vast literature. Nonetheless the views of selected gender-theorists provide intriguing possibilities. The notion of "Concept Amelioration" (Haslanger), for example could prove helpful for the Conceptual Question. And, the "Multiple-Meaning Position" (Bettcher), while aiming to establish the non-marginal status of Trans-women and Trans-men as Women and Men, also allows the positing a new ontological phenotype—Trans-women/Trans-men—an idea elaborated upon in the remaining chapters of the book.

Keywords Gender-theorists · "Multiple-meaning position" · New ontological phenotype: Trans-men/Trans-women · "Concept amelioration"

There is, as one would expect, a sizeable literature on current views of gender, most prominently from feminist philosophers. The relations of gender, sex, biology, and society are examined with much interest, creatively and critically, in a number of these works, both regarding Trans and Cis individuals. For example, there is a group of essentialists who

L. A. W. Brakel, *Investigations into the Trans Self and Moore's Paradox*,
https://doi.org/10.1007/978-3-030-44645-1_4

consider "gender" a real and natural kind, marked by common core activities, attributes, and concerns; these due to social and/or biological features shared by "women" on the one hand, and "men" on the other. Critics of this view have claimed that this type of essentialism is untenable given that most real and natural kind essentialists do not give sufficient weight to historical and contextual considerations. Among the opponents to real and natural kind essentialist, one can find widely diverse thinkers. There are those who refine the above essentialist notions, considering the categories of "gender" and particularly "woman" as a real, but not a natural kind. Instead, for this group of critics, these categories are socially constructed, and even culturally determined by a mainstream androcentric culture. More radically, some non-essentialist theorists consider the very idea of a coherent category of "woman" a myth of the patriarchy. Cross cutting all of these views—including essentialists, non-essentialists, cultural feminists, and post-structuralists[1]—there are theorists who propose that "gender" cannot be understood materially at all, but instead performatively. This view, can be characterized by a slogan from one of its most eminent developers, Judith Butler (1990, p. 25) who states that "gender is doing," Butler contends further (1990, p. 136): "That the gendered body is performative suggests that it has no ontological status apart from the various acts which constitute its reality."[2]

Important and compelling, these issues are themselves orthogonal to the current project. This owes to something rather surprising to me. Within this wide, deep, dense, scholarly, and biographical literature, there is a dearth of material about the nature of *Self*, both in general terms, and for Trans individuals; and particularly with regard to the epistemological and ontological questions central to the current project—viz, How and what does a Trans person, like our emblematic G→B know leading to a transition? and What ontological aspect of G→B's *Self* is operative in knowing and then effecting this? Thus, even in the accounts to be taken up in this section—a section which has among its goals to

[1]Within these named categories, there is much overlap. Many of these accounts do not form a unified category and are difficult to classify.

[2]Butler quotations are from Alcoff (2006, p. 157); Also, my brief review in the above paragraph owes to Alcoff (2006, Chapter 6, particularly pp. 151, 157), Bach (2012, pp. 231–232), and Haslanger (2000, p. 37).

explore the *Self* views of contemporary philosophical gender-theorists—I must point out that most approach gender from angles very different from those motivating this monograph. To highlight the fact that indeed, my quest to resolve foundational philosophical problems is at odds with the aims of many (most) prominent philosophers and gender-theorists, I will provide (in footnote 3 just below) a list of several articles, chapters, and books that do not engage with the fundamental epistemological and ontological questions regarding Trans persons and the *Self*.[3,4]

Nonetheless, with apologies for the brevity, and for the seemingly random selection of the accounts I do report upon,[5] the substance of this section will consist of discussing a number of interesting, original, and important philosophical views on gender, ending with two accounts which do manifestly discuss the *Self* in ways that can be shown to intersect with the interests of this project. Then, at this section's end there will be a bridge to biological matters, followed by Chapter 5's original proper-function account of gender and Trans-gender.

[3]Here is the list: Carson, A. (2016). Sex, biological functions and social norms: A simple constructivist theory of sex. *Nordic Journal of Feminist and Gender Research, 24,* 18–29; Haslanger, S. (2017). Social construction: Gender and other social categories. In A. Garry, S. Kahder, & A. Stone (Eds.), *Routledge Companion to Feminist Philosophy.* New York: Routledge; Yarbrough, E. (2018). *Transgender Mental Health.* Washington, DC: American Psychiatric Association Publishing; Green, J. (2004). *Becoming a Visible Man.* Nashville: Vanderbilt University Press; Nutt, A. *Becoming Nicole: The Transformation of an American Family.* New York: Random House; and Barres, B. (2018, Fall). Deciding to change: An excerpt from the autobiography of a transgender scientist. *Plus, the Stanford Medicine Newsletter,* 34–35, 41.

[4]Note that one of eleven articles in Laurie Shrage's (2006) edited collection, *You've Changed: Sex Reassignment and Personal Identity.* New York: Oxford University Press, does deal with *Self* in ways relevant to this project. That article, taken up below, is Christine Overall's "Sex/gender transitions and life-changing aspirations," Chapter 1. Another chapter in Shrage's volume (Chapter 6), Talia Mae Bettcher's "Trans Identities and First-Person Authority," discusses Self, but from an ethical legal viewpoint that is quite apart from the issues explored in the current work. Finally, Alcoff's *Visible Identities: Race, Gender, and the Self,* does, as the title indicates, indeed consider aspects of the Self, but again from a very different vantage point from the questions about the *Self* explored here. For instance, there is a section (pp. 113–125) on knowledge of *Self*/Other. Here various accounts are developed which suggest that the *Self* is derived from interactions with the Other.

[5]The accounts I have selected are salient with respect to the issues I will later elaborate, offering extremely valuable insights.

1 Gender as a Natural Kind: Historical, Not Strictly Biological, Not Strictly Social

Theodore Bach (2012) argues that gender is a natural kind, whereby natural kinds are real "...ontological structures that support successful inductive practice" (p. 240). For Bach, gender is moreover a historical kind, one fulfilling the same sort of teleofunctioning for the "gender system's" selective fitness that the teleofunctioning of biological natural kinds fulfill for the selective fitness of a biological system. Thus, the function of the heart is to pump blood. Individuals inheriting a heart from progenitors with hearts will have had selective advantages over other historical alternatives. Hence, they will produce more offspring, all with inherited hearts, and all stabilizing the cardiovascular system, itself a product of historical selective success. Various problems can interfere with producing a normal heart, but defective hearts that fail to pump blood are hearts nonetheless because "...they have the right historical relation to other hearts" (p. 244).

So, how does Bach apply this to gender. First (p. 246) he explains that on his view "...being a woman or a man is not a matter of possessing some set of biological or psychological properties, as these are merely probabilistic indicators...Nor for the same reason, is it to occupy a certain social position..." He continues (p. 247): "My claim is...several processes that, through their causal interdependence, have led to the selection, maintenance, and ongoing reproduction of gender systems." After noting in a footnote (number 10 in Chapter 3) that there are no a priori reasons that gender systems should be or stay binary, Bach goes on (p. 247) to characterize the current binary gender system as "...homeostatic,...its various components have achieved a stability that is resistant, though not impervious to change, ...and it certainly does not follow that the system is desirable or ethical." Bach next outlines seven components of the current binary gender system that provide the selective fitness for that system. These are (pp. 247–248): (1) Binary Sexual Categories; (2) Conceptual Gender Dualism; (3) Gender Identity; (4) Binary Gender Socialization Practices; (5) Social and Legal Institutions; (6) Binary Gendered Artifacts; and (7) Binary Gender Roles. Then, elaborating on these binary gender roles, Bach offers (pp. 254–255) candidate roles for men and women—roles which may have, and may continue to contribute to the selective fitness of the current binary gender system. These are: (1) Social hierarchy; (2) Division of Labor; (3) Personal and Interpersonal Characteristics; and (4) Body Management. Finally, he holds (p. 255), "We can assume that ancestral males and ancestral females exhibited a range of social and behavioral traits. Among these competing gender role cultural variants, sexed individuals who exemplified properties 1-4...acted to stabilize that system [i.e., the current binary gender system] more than individuals who did not exemplify these properties. If these gender roles were culturally reproduced on account of this stabilizing effect, we can conclude they were culturally selected for."

1.1 Implications of Bach's View for the Trans Foundational Questions

It is obvious that Trans persons, even those who are the focus of this project—namely Trans persons who do embrace the mainstream binary categorization of gender (male or female), but who feel they are the "wrong" one—do not fit most (and maybe none) of the specifications in Bach's list of seven component parts constituting what he deems the current stable binary gender system. Nor do Trans persons fit his list of four gender role stabilizing properties. But these are not particular problems for Bach's view. He can explain and accommodate these differences in terms of an analogy to the cardiovascular system: unusual gender roles are to the regular binary gender system as cardiac anomalies or vascular peculiarities are to the ordinary normal cardiovascular system. Hearts that do not function properly by pumping blood in the ordinary or regular way, are still hearts as long as they have been inherited from hearted individuals as part of stabilizing the selective fitness enhancing cardiovascular system. So, on Bach's view, Trans-women and Trans-men can be, and indeed are women and men respectively. (This view will be elaborated upon, and enhanced as it is approached from a different direction. See especially the subsection on Bettcher's work on the "Multiple-Meaning View of Gender" below).

But Marion Godman raises a potentially troubling issue about Bach's account in her 2018 article, "Gender as a historical kind: a tale of two genders." First, summarizing Bach, Godman states (p. 32): "He supposes that…for a gender system: the advantages originally conferred by certain traits belonging to a gender system is also the reason why these traits are reproduced. It is the favorable effect of gendered traits within a cultural gender system in the past that causes descendants of that culture to continue to possess them." Then she comments (p. 32, footnote 11 in Chapter 1), "…it is not clear whether it is cultural *group* selection that he has in mind or if it is also an advantage to the individual trait bearer's reproductive fitness" [Godman's italics]. While I think Bach has pointedly stated that that it is selective fitness for the binary gender system itself, i.e., neither group selection, nor the individual, Godman's worry and her comment are instrumental in my derivation of a biological proper-function account for gender and Trans gender, to be developed in the next chapter, Chapter 5.

As a final note about Bach's work, it is worthwhile to point out, that the foundational philosophical *Self* questions—ontological, epistemological, and conceptual—highlighted by the Trans phenomenon, are not the focus of his investigations of gender.

2 Bettcher's Multiple-Meaning View of Gender; Haslanger's Concept Amelioration Project

Talia Mae Bettcher (2012, p. 234) argues that the "Wrong Body View"—an account of Trans persons in which "real" females are trapped inside male bodies, and "real" males are trapped inside female bodies—is a very flawed view. Equally flawed, according to Bettcher (p. 234), is what she terms "The Transgender Model"—characterized by the claim that "…it is not trans people who are the problem but society itself." For Bettcher (pp. 234–235) the "Wrong Body View" reifies and biologically naturalizes gender identities, in a way not endorsed by all Trans persons, and is thereby over generalized. Similarly, overgeneralization is a problem for the "Transgender Model" which cannot be endorsed by Trans persons who do not see themselves as "beyond the binary"—but as part of the binary. Indeed, the group at the center of this project is the very group of Trans persons who do see themselves as binary, but mismatched with their natal gender.

Continuing her critique of both views, Bettcher (p. 236) points out that when "woman" is defined as "adult female human being" and "man" as "adult male human being," proponents of one view consider these differences biological, whereas those favoring the other account argue that "woman"…"picks out a *social* kind, role, or status" [Bettcher's italics]. The diagnosis Bettcher applies is that each of these views is but a "single-meaning position" for a concept that requires more than one position. She proposes instead (pp. 240–244) "the multiple-meaning position." Claiming that one of the advantages of her multiple-meaning position is that it takes cultural context into account, Bettcher first cites several experiences of Trans persons. She states (p. 240): "I know many trans-women…who are content with their 'male genitalia.' However, many do not consider them *male* genitalia in the first place, but the sort of genitalia congruent with transgender femaleness. Similarly, I know many trans-men who have no interest in phalloplasty and who consider their genitalia (transgender) *male* genitalia" [Bettcher's italics]. Comprehending those situations, Bettcher explains (p. 240) that "…what happens is that the social meaning commonly associated with a body part is, in a subcultural context, completely changed."

An outgrowth of the socio-cultural advantages of her view, Bettcher's "multiple-meaning position" can also function in a salutatory fashion with respect to the concepts, "woman" and "man," and their extension. Given, that the extension of a concept is standardly understood as describing the members or objects that fall under that concept's range, Bettcher's "multiple-meaning position" has allowed an expansion of the concepts, "man" and "woman," extending their extensions.

This is much in keeping with the sort of concept-evolution that Sally Haslanger (2000, p. 33) recommends in her own project—a conceptual inquiry into gender. Convincingly explaining the idea of expanding a concept's extension, Haslanger (2006, p. 105) offers a more familiar case, discussing the concept of "parents." Says Haslanger, "…we once took it to be an essential feature of parents that their children were biologically related to them, but we have come to regard this as an empirical generalization based on a limited survey of cases that does not hold necessarily. This is a change in our conceptual knowledge." Moreover, Haslanger's (2000, 2006) suggestions for concept revision go beyond this sort of social reality inflected expansion—namely, toward what she terms concept amelioration. She explains (2006, p. 95): "Amelioration projects… begin by asking: What is the point of having the concept in question…" Then asks further (2000, p. 33), "…if our concepts are serving our (legitimate) purposes; [and] if not, what concepts would serve these purposes better?" Again, with respect to her own work, arguing for changes in the concepts of gender (and race) to reflect (and enable) political reform, Haslanger concludes (2000, p. 52), "…rather than worrying, 'what is gender, really?' or 'what is race, really?' I think we should begin by asking (both in the theoretical and political sense) what, if anything, we want them to be."

Returning to Bettcher's work, we should ask what exactly would be entailed in accepting Bettcher's "Multiple-meaning position?" She describes the process in two simple steps. "First, 'trans woman' is taken as a basic expression, not as a marginalizing qualification of the dominant meaning of 'woman' (p. 240)…The second step (p. 241) is that being a trans woman is a sufficient condition for being a woman. 'Woman' is then taken to apply to *both* trans and non-trans-women…" [Bettcher's italics].

2.1 Comments on Bettcher's "Multiple-Meaning Position" and Haslanger's "Concept Amelioration Project"

Bettcher's "Multiple-Meaning Position" is both interesting and important. Although it is directly set forth in a linear fashion, it also consists of two less obvious interlocking aspects. One aspect reveals a subtle and coherent logical analogy and, is in this way quite aesthetically pleasing: Thus, just as small squares, large squares, red squares, and blue squares, are all squares, Trans-women, and Cis-women too, are all women. The other aspect is an ontological issue that Bettcher's work hints at, and thereby introduces: namely, that within the ordinary mainstream binary gender system, consisting now of Men and Women, there is the possibility of two additional ontological phenotype categories—Trans-men; and Trans-women. Haslanger's position on concept expansion and amelioration too might offer strong support to these new ontological phenotypes. Note however, that neither Bettcher's work nor Haslanger's addresses the concept of *Self*—and how it might be extended or improved to accommodate the Trans experience.

Of course, I must admit, that despite my direct and intense focus on that concept exactly, I too fall short here. And yet, the idea of an expanded ontological phenotype—one that includes Trans-men and Trans-women within the mainstream gender binary—will be more central in the remaining chapters of this book. In Chapter 5 these two new ontological phenotype categories will have a role in the proper-function analysis of gender and Trans-gender that I advance; then in Chapters 6 and 7 various sorts of evidence for (and against) these two new phenotypic categories will be accrued—Chapter 6 presenting brain findings on the Trans phenomenon; and Chapter 7 comprising several empirical Experimental Philosophy experiments derived from classic Thought Experiments on the *Self*.

2.2 Implications of "The Multiple-Meaning Position" and Haslanger's "Concept Amelioration Project" for the Issues of *Self* Raised by Trans Experience

The ontological, epistemological, and conceptual questions regarding the *Self* as they relate to Trans persons are not matters taken up by Bettcher in her chapter, nor by Haslanger in the two articles cited. Bettcher does provide, instead, a view closely aligned with Haslanger's work, whereby the extension of the gender concepts of men and women could be usefully expanded, ameliorated so that these categories would no longer be singularly fixed—not by biology, nor social convention, nor by the negation of either biology or social convention.

3 A Dispositional Account of Gender

Jennifer McKitrick (2015, pp. 2575–2589) presents a view of gender that attempts to bridge social construction accounts thereof, instead offering a more psychological subjective approach. With respect to gender as socially constructed, McKitrick references a few thinkers, among whom are Charlotte Witt (2011) and Sally Haslanger (2000).[6] McKitrick (p. 2576) summarizes Witt's notion of gender(s) as "…social positions with bifurcated social norms that cluster around a socially mediated reproductive function." And then she quotes Haslanger (2000, p. 39): "S is a woman iff S is systematically subordinated along some dimension (economic, political, social, etc.), and S is marked as a target for this treatment by…bodily features presumed to be evidence of a female's biological role in reproduction." Finding views such as these wanting with respect to a particular subject's experience, McKitrick suggests that the gender-as-social-construction views are more about gender role than gender identity.

Then, because she seeks an account that is more about gender identity, McKitrick turns to Butler's notion of gender as performance. McKitrick (p. 2577) states: "According to Butler, patterns of dress, posture, and speech are not expressions of an inner identity, but are instead constitutive of being gendered." In fact, McKitrick continues (p. 2580), Butler provides "…an alternative to a conception of gender as an essential innate truth about a person and gendered behavior as an outward expression of an inner gendered self… [Moreover], Butler turns this notion on its head and construes this inner self as a consequence of, or a construction out of, these behaviors rather than their cause."

But, as pleasing as this view is to anti-essentialists (of which McKitrick is one), McKitrick recognizes that Butler's view cannot account well for the profound gender mismatch feelings of Trans individuals, both those with and without gender dysphoria. Nor can Butler's position accommodate instances of gender dysphoria in other populations.

---

[6]Witt, C. (2011). *The metaphysics of gender* (pp. 32, 40). New York: Oxford University Press; Haslanger, S. (2000) is referenced above and in the Bibliography.

For Butler, according to McKitrick (p. 2578), "There is no inner, other-gendered self to conflict with one's socially sanctioned gender role." So McKitrick attempts what she considers a correction, involving dispositions (pp. 2578–2579): "I think it is better to think of genders as *dispositions* to behave in certain ways, dispositions which are not always manifest" [McKitrick's italics]. She elaborates (p. 2580), first describing gender dispositions as "massively multi-tracked" and occurring as a "cluster of dispositions," and then listing various circumstances as triggers for gendered behavior. Among these McKitrick includes (p. 2580) "…shopping for clothes, getting a haircut, being insulted, being asked what you want for Christmas, registering for classes, etc."

3.1 Problems with This Dispositional View and Implications for the Trans Questions

The McKitrick solution does not appear to work. The author herself fears that the anti-essentialists will find her view too essentialist in that they'll contend she envisions some sort of innate feminine disposition to behave in feminine ways. No, McKitirck says (p. 2580), "It is consistent with my approach that dispositions for gendered behavior are products of socialization, are had in varying degrees of intensity, and could change over time." Now, I'm not sure that this argument will assuage the anti-essentialists, but one thing is clear. As with Butler's view in which gendered behaviors form the gendered-*Self*, rather than the *Self* constituting the gendered behaviors; with McKitrick's dispositional view we have what amounts only to an addition, not much of a correction. Namely, insofar as the disposition to act in a gendered way, along with acting so gendered, forms the core *Self*, and not vice versa, we have gained no traction on what ontological aspect of the *Self* drives the "multi-tracked, cluster of dispositions" in the first place—neither for Cis, much less Trans individuals.

But McKintick does takes up the issue of *Self*, and seriously so, as does Christine Overall, whose account follows.

4 Aspirational View of Gender Identity

Christine Overall (2006) in a chapter called, "Sex/gender transition and life-changing aspirations" begins by asking the following questions about the metaphysics of gender transition: "Who or what is changed, and who or what remains the same?" Seemingly similar to problematic questions motivating the current work—namely: What ontological aspect of the *Self* knows that a gender transition is desired and needed? What is it that is epistemologically known to that *Self* aspect?—the answers Overall provides to her questions suggest that the similarity is at surface level, not structurally deep.

Like Bettcher's (2012) "Multiple-meaning position" on gender discussed above, Overall (2006, p. 12) disputes both "masquerade" models of gender transitions whereby "...either (1) the 'true' person is thought to be the individual manifested...by the original sex/gender status, and the transitions is, metaphorically...a mask that conceals the true sex/gender, or (2) the 'true' person is thought to have been hidden behind a metaphorical mask of the wrong sex/gender, and the individual's true sex/gender is revealed via the process of transition." She locates the problem with both of these views in the notion of "...a reified self that constitutes the core of the individual and does not change during the ostensible transition."

Overall's alternative account—reminiscent of Hume's eliminative view of any singular *Self* investigating the perceptual particulars he encounters when he so investigates—is as follows: While acknowledging there are causes involved in producing the idea that one is a different gender than the one signaled bodily, and that these causes may be contributed to by biological factors, Overall (p. 18) does not seem at all exercised by the vexing epistemological or ontological questions at the heart of this monograph. She says, "...when trans persons make claims about their sex/gender, they...are offering an *interpretation* of their embodied being..." [Overall's italics] Next, given that for Overall (p. 16) "...gender is thoroughly social," she goes on to explain (p. 19) that, "The metaphysical problem of identity in sex/gender transitions arises from assuming that these transitions must necessarily be qualitatively different from other major changes in people's lives..." Indeed, Overall asserts (p. 19),

"Sex/gender transition is best understood…by analogy to other life-changing and life-enhancing aspirations for personal transformation and self-realization." She lists other sorts of aspirational changes—becoming an immigrant, changing religions, taking on an important new role such as parenthood—but does agree (p. 19) that, "The person who undergoes a sex/gender transition is someone who seeks a radical transformation of and in her or his sex/gender project."

4.1 Problems with the Aspirational View; Implications for the *Self* Questions Highlighted by the Trans Experience

Overall (p. 23) anticipates exactly what critics will (and do) find problematic in her account, as she voices the skeptic's questions: "Where do the aspirations of trans persons come from? How can they be explained?" And She answers (p. 23): "…my point is that aspirations for sex/gender transitions are not necessarily different in kind from other deeply felt and long-held aspirations." She then notes a network of biological, social, and personal historical factors all influencing persons, concluding (p. 23), "In some people, this network…contributes to a desire to be a musician; in some an eagerness to be a parent; in some, they produce spiritual aspirations; and in some, they produce an aspiration to be a person of a different sex/gender." And yet, I find myself aligned with the skeptics, for I too must ask (as I have often in this monograph): From what *Self* or aspect thereof does the very radical aspiration to change gender arise? This question leads, almost inexorably, to the original questions animating this project—How do persons know that gender assigned natally is "wrong," and not to be endorsed? What do they know? And, what part of the *Self* does that knowing?

* * *

Taken together, over the last two chapters, many modern philosophical accounts of *Self*, including those of the feminist gender-theorists, were presented, reviewed, and commented upon. Like the classic philosophers on the *Self* discussed earlier, these many modern philosophical views have been less than successful in providing answers to the three puzzling foundational philosophical questions motivating this project: (1) The epistemological query as to how and what Trans persons know regarding their natal gender's "wrongness"? (2) What is the ontological nature of the Trans person's *Self* (or aspect thereof) engaged in this knowing? and (3) Do these matters suggest that a conceptual change to the concept of *Self* in general is warranted?

These philosophical investigations, however, while disappointing in that regard, have opened the way to a different set of issues to consider. Thus, while no longer directly focusing on the search for answers to the three foundational philosophical questions, the remaining chapters of this work will explore questions that can, in principle, be more successfully addressed. Thus, three different tacks, original and unusual (especially in their co-appearance in a single monograph) will be taken. First, in Chapter 5 I will propose an evolutionary based proper-function account of gender, including Trans-gender. Next, Chapter 6 will explore issues involving the possibility of brain-based *Self* differences in Trans persons. Many recent (1995–2018) brain-based studies are reviewed—some of which make etiological claims (i.e., brain changes are the causal agents of Trans-ness), and others holding that epigenetic brain findings, secondary to living a Trans life, best account for the Trans phenomenon. In both Chapters 5 and 6 the possibility of a new ontological phenotypic category—Trans-men; Trans-women—still within the mainstream binary gender system, will be explored. Fish will play a surprising role in both Chapters 5 and 6, providing proof-of-concept type evidence for some of the hypotheses posited. Then, the penultimate chapter, Chapter 7, presents a series of empirical Experimental Philosophy (X-Phi) studies designed to accrue more evidence for or against these hypotheses. Adapted from several classic Thought Experiments on *Self* (discussed at some length in Chapter 3), these X-Phi experiments also are shown to have potential pragmatic application. In this way, our frustratingly unanswered questions might paradoxically lead to discovery in a different direction. Such is the way, when lucky, in both philosophy and science.

Bibliography

Alcoff, L. (2006). *Visible identities: Race, gender, and the self.* New York: Oxford University Press.

Bach, T. (2012). Gender is a natural kind with a historical essence. *Ethics, 122,* 231–272.

Bettcher, T. (2012). Trans-women and the meaning of "woman". In N. Power, R. Halwani, & A. Soble (Eds.), *The philosophy of sex* (Chapter 14, pp. 233–249). New York: Rowman & Littlefield.

Butler, J. (1990). *Gender trouble: Feminism and the subversion of identity.* New York: Routledge.

Godman, M. (2018). Gender as a historical kind: A tale of two genders. *Biological Philosophy, 33,* 21–36.

Haslanger, S. (2000). Gender and race: (What) are they? (what) do we want them to be? *Nous, 34,* 31–55.

Haslanger, S. (2006). Philosophical analysis and social kinds: Sally Haslanger and Jennifere Saul. Part I. Haslanger: What good are out intuitions? *Proceedings of the Aristotlian Society, 106,* 89–118.

McKitrick, J. (2015). A dispositional account of gender. *Philosophical Studies, 172,* 2575–2589.

Overall, C. (2006). Sex/gender transition and life-changing aspirations, In L. Shrage (Ed.), *You've changed: Sex reassignment and personal identity* (Chapter 1, pp. 11–27). New York: Oxford University Press.

Witt, C. (2011). *The Metaphysics of Gender.* New York: Oxford University Press.

Bibliography

Part III

Biological Considerations

5

A Proper-Function Account of Gender and Trans-Gender: A Bridge From Philosophical Views to Biological Matters (Including Fish!)

Abstract Several of the accounts in the previous chapters feature evolutionary considerations. These, in conjunction with the "Multiple-Meaning Position"—Bettcher's work which argues for non-marginal status for Trans-women as Women, and Trans-men as Men—provide support for the original proper-function account of gender, and in particular, Trans-gender derived and developed in this chapter. This account entails a complex argument: It takes its structure from the work of Ruth Millikan, followed by my extension of her work in the direction of a-rational categories, capable of wider biological application. This is all spelled out in detail in the chapter. Finally, sex-changing fish provide "proof-of-concept" evidence for the proper-function argument.

Keywords Evolutionary proper-function account of gender/Trans-gender · New ontological phenotype: Trans-men/Trans-women · Proof-of-concept: sex-changing fish

L. A. W. Brakel, *Investigations into the Trans Self and Moore's Paradox*,
https://doi.org/10.1007/978-3-030-44645-1_5

1 Brief Introduction and Orientation

Several of the modern philosophical views discussed in the last two chapters have had an impact on the proper-function evolutionary selective fitness account of gender and Trans-gender that I will develop in this chapter. Additionally, observational research on certain fish species that change their sex, and on other fish species that modify their morphological gender will be described, affording proof-of-concept type support to the proper-function analysis.

2 A Proper-Function Account for Gender and the Trans Experience

The proper-function evolutionary account I will present is complex and will require explanation in a number of steps. First, as the view derives from work done by Ruth Millikan, particularly several of her 1993 essays in *White Queen Psychology and Other Essays for Alice*, let me begin with a brief review of Millikan's general notion of what she terms "Proper-function under Normal Conditions." The proper-function of some process, or system, or organ, X, must include X's adequate function under so called "Normal Conditions." Normal Conditions are those in which the organism, Y, having inherited the adequately functioning X, is aided in proliferating (i.e. aided in its selective fitness success) due to having inherited adequately functioning X. Fine, but now let's take a complicating situation. Consider a human sweat gland that is adequately functioning, secreting sweat, but is secreting sweat in response to a message of overheating due to a tumor at the temperature sensing area of the brain. This healthy sweat gland, although itself well-functioning, is not functioning properly under Normal Conditions because the tumor-driven message of overheating constitutes an abNormal Condition. The sweat gland by secreting sweat can only be functioning properly under Normal Conditions when the temperature sensing area correctly reports that the body is overheating. Only under these Normal Conditions would an inherited well-functioning sweat gland contribute to enhanced selective fitness success for the human inheriting it.

Now for the next step, let's bring this sort of analysis to something more psychological, more mind/brain relevant—beliefs and belief-generating-mechanisms in humans. Again, mechanisms that are properly functioning under Normal conditions are those mechanisms that when inherited enhance selective fitness success. Thus, with respect to beliefs and belief-generating-mechanisms, it is fitness normativity, rather than "rational normativity" that allows for: (a) contentful (representational) mentation; (b) a-rational,[1] as well as rational mentation; (c) well-functioning belief-generating mechanisms that produce beliefs—both true and not-true. One might question the notion that not-true (i.e., false beliefs) can be adequately functional. But, much as is the case for the healthy sweat gland producing sweat in response to a tumor-driven signal, false beliefs can still be considered adequately functioning, given that there are sufficient true beliefs generated. Why? And, how does this work? Here are the answers: Yes, it is the case that only the true beliefs produced are properly functioning under Normal conditions; and similarly, it is only the true beliefs that can contribute to selective fitness success in organisms inheriting belief-production capacity. But the not-true or false beliefs generated may well reflect adequate belief-forming-mechanisms, in that they operate under abNormal conditions—either external (e.g., limited visibility such that a cow is taken for a horse) or internal (e.g., a milk deprived person mistaking a horse for a cow). Pertinent here is an important point from Millikan (1993, p. 91): "Many biological devices perform proper-function [under Normal Conditions] not on average, but just often enough..."[2]

And now, for one final twist before getting to gender issues, let's examine frog "beliefs" and categories. It is reported that frogs swallow indiscriminately, on reflex, small black objects—both bugs and metal pellets. With selective fitness success normativity operative as that which can determine representational content, it follows that the frog category

[1]More on a-rationality just below.

[2]For a more complete description and explanation of these matters, in particular, how replacing rational normativity with fitness normativity allows a-rational mentation to be determinatively contentful, see Brakel (2009, Chapter 5, pp. 63–83, particularly pp. 70–83). "Representational a-rational thinking: a proper-function account of phantasy and wish."

"black objects to ingest" can best be considered as a properly functioning, a-rational category[3]—BUG = black object to ingest—rather than a frequently defective attempt at a rational category, often misrepresenting bugs. Why, and how can that be so? Frogs reacting to the inclusive BUG category by swallowing and then ingesting bugs when bugs are present, demonstrate proper-function under Normal Conditions. On the other hand, frog swallowing and ingesting items in the BUG category that occur when only metal pellets are present, are still demonstrating swallowings and ingestings that are functioning adequately, but now they are functioning under the abNormal Condition of no bugs present, only metal pellets. Certainly, eating bugs contributes to frog nutrition, and this in turn aids in frog reproduction and thereby selective fitness success. Eating non-nutritive metallic pellets, of course, does neither. But it is at least plausible (likely), that the a-rational inclusive category, BUG, allows more bugs to be eaten with fewer bugs missed, than would a narrower rational category. Thus, it is posited that the more inclusive a-rational category might itself promote selective fitness success (provided minimal toxicity in pellet swallowing and ingesting). In that case, while each black metal pellet swallowed constitutes an abNormal Condition, the a-rational inclusive category might on the whole serve as an overall Normal Condition properly functioning category. (Ruth Millikan's remark about "just often enough" is pertinent here.)

Now back to humans and the gender matters at issue. Could a similar proper-function evolutionary account work for gender, and particularly for Trans-gender persons?[4] The gender conditions considered properly functioning under Normal Conditions would map Woman onto females, specifically those who are fertile; and Man onto males, specifically those who produce viable sperm. But an inclusive gender category (analog to the frogs' category, BUG) would open the WOMAN category to various sorts of non-fertile females—those that are menopausal, infertile or

---

[3]Take note that this category is a-rational, not irrational. See Brakel (2009), particularly Chapters 1, 5, and 7 for much more on this distinction.

[4]Clearly, the proper-function analysis to be developed here is quite distinct from that of Bach described above (Chapter 4). Bach is dealing with the mainstream gender system as a system; its ongoing proper-function reflective of its stability and success as a system.

not-wanting-kids, pre-menstrual, and importantly male-to-female Trans-women.[5] (One can do the same for MAN and various non-fertile males.) The inclusive categories, and their formation, would be adequately and properly functioning; but only properly functioning under Normal Conditions when WOMAN referred to fertile females, and MAN picked out fertile males. These conditions are the only ones called "Normal" because only these contribute to selective reproductive fitness. In the case of any and all of the non-fertile sorts, although functioning properly, and properly categorized as WOMAN (or MAN), they would be functioning under so called abNormal Conditions.

But even as that is made clear, there are reasons (at two inter-related levels) arising from this analysis that serve to question the seemingly straightforward conclusions given just above: First, would an analogy with frogs, with their enhanced fitness success potential given the more inclusive BUG category, hold true with the MAN and WOMAN inclusive categories? Put simply, could the more inclusive gender category—even with the many instances of abNormal Conditions, but given the need for only intermittent proper-function under Normal Conditions—prove on the whole to constitute a fitness success enhancing properly functioning Normal Condition category? In other words, since a properly functioning Normal category could entail some (perhaps many) abNormal instances, would humans, as a species, be better off embracing this sort of inclusive a-rational category?[6]

Second, looking now at individuals, would one's reproductive fitness success be enhanced by transitioning, if gender transitioning were wanted and needed? As Sarah Buss's (2012, p. 660) work underscored (described in Chapter 7), real agency entails actions that accord with "minimal human flourishing"—flourishing that we can surmise would be impossible unless one could feel gender-aligned. Taking this a step further, if transitioning allowed Trans persons to flourish, transitioning for such

[5]This account squares well with Bettcher's (2012) work (see Chapter 4), in which Trans-women are women, sufficiently so, and not marginalized.

[6]Certainly, various philosophers and social theorists have suggested this. See Chapter 4 on Bettcher (2012) and particularly Hasslanger (2000, 2006).

persons could (perhaps paradoxically) promote evolutionary fitness success in that only under these conditions could a person have the wherewithal to participate in the care and raising of children having some percentage their own genetic material.[7] Indeed, in some instances Trans persons have reproduced (and/or will do so)—this, either before transitioning, or when transitioning does not include hormonal changes and genital re-assignment. My own view of agency (summarized in Chapter 3) accords here with Buss's. Persons transitioning can be seen as acting intentionally, responding not just to Why questions seeking reasons; but to the biological Why questions too—where the answers are actions with the potential for enhancing evolutionary fitness success.

2.1 Fish

As promised, it is (finally) time to examine fish. It is certainly the case that when fish who are members of a sex-switching species undergo sexual/gender transitions, they do so only when social conditions favorable to such switches are perceived as such. And, it is just as certain that favorable in this context means that sex-switching can potentially enhance the reproductive fitness of the sex-switched individual, thus enabling more success in passing along that fish's genes to future generations. Here is an example from Godwin (2010, p. 206): "Bluebanded gobies typically live in groups consisting of a behaviorally dominant male and several females in nature. Removal of the male from a social group leads to a rapid gonadal, morphological, and behavioral changes in the largest female

[7]A New York Times article (Styles Section, p. 1), January 17, 2019 features two homosexual penguins (a non-reproductive couple) doing a good job of parenting—tending to an egg through its hatching, then feeding, protecting, teaching—in all respects raising the chick. Although in this case, there was not a close genetic connection, there well could have been. The "grandmother hypothesis," in which older non-fertile women help care for those genetically related to them, is relevant here too—promoting selective fitness success for the grandmothers' genetic material. (See Hawkes, K., O'Connell, J., Jones, N., Alvarez, H., & Charnov, E. [1998]. "Grandmothering, menopause, and the evolution of human life history." *Proceedings of the National Academy of Sciences, 95*, 1336–1339). In between penguins and humans, we have non-fertile grandmother orcas, who are subjects of a recent article, "Post reproductive killer whale grandmothers improve the survival of their grand offspring." (*Proceedings of the National Academy of Sciences.* December 2019, 201903844. https://doi.org/10.1073/pnas.1903844116).

as she undergoes protogynous [female-to-male] sex change." This former female fish, now as the sole male, will be a most successful breeder, launching his genetic material to future generations.

Much more will be said about the behavioral and brain mechanisms operative in this sort of sex-switching, and the opposite type, protoandry [male-to-female], in Chapter 6, where in addition to reviewing recent brain studies on Trans vs. Cis humans, there will be a section on the brains of sex-switching fish. But before we leave this chapter, let me explain why fish have been included here too. I propose that we regard these sex-changing fish as analogous in important ways to human Trans persons. As such, the straightforward claim is that these sex-switching fish, with their enhanced selective reproductive success after sex-change, can provide a proof-of-concept anchor for the evolutionary proper-function account for Trans persons I have just advanced.

Furthermore, and more speculatively, I wonder if the sunfish "satellite male morph" mentioned in the Introduction Chapter could provide a further analogy. Recall that these "satellite male morphs" are fish that are male only in their gonads and gametes, but female in their morphology, behavior, and likely in their brains, including gene expression and epigenetic influences therein.[8] Could we consider these satellite fish not only "male morphs" but also "trans-females," and in this way quite like some of the Trans-women Bettcher describes—WOMEN, Trans-women, who do not desire change from their natal (male) genitals and gonads?[9] This is an exciting parallel and it leads to another: Could classifying a sunfish satellite morph as a trans-female fish allow for (perhaps give evidence toward) the possibility of a separate ontological phenotype—Trans-man; Trans-woman—with this new phenotypic category remaining within the mainstream binary gender/sexual system?

This idea will be explored more fully in Chapter 7, with empirical Experimental Philosophy (X-Phi) adaptations of classic *Self* thought

[8]Godwin (2010, p. 210) explains that the neuroendocrine brain findings are not yet complete, but that data from related species indicate that on examining their brains, the satellite morphs, while somewhere in between parental males and females, may be closer to the females.

[9]The evolutionary proper-function analysis pertains here both to fish and humans. Just as these male-morp/trans-female fish increase their reproductive success, some of the Trans-women (WOMEN) Bettcher describes express their agential human flourishing by choosing to perpetuate their genes through unaltered male gametes.

experiments, and also in Chapter 6. Chapter 6, which comprises a review of neuroscience brain studies with Trans persons (and Cis-controls) done in the last 25 years, also includes a section on brain and behavioral studies on sex-changing fish—particularly their social situations, perceptions thereof, and the potential relevance for the human Trans phenomenon.

Bibliography

Bettcher, T. (2012). Trans-women and the meaning of "woman". In N. Power, R. Halwani, & A. Soble (Eds.), *The philosophy of sex* (chapter 14, pp. 233–249). New York: Rowman & Litterfield.

Brakel, L. A. W. (2009). *Philosophy, psychoanalysis, and the a-rational mind.* Oxford: Oxford University Press.

Buss, S. (2012). Autonomous action: Self determination in the passive mode. *Ethics, 122,* 647–691.

Godwin, J. (2010). Neuroendocrinology of sexual plasticity in teleost fishes. *Frontiers in Neuroendocrinology, 31,* 203–216.

Millikan, R. (1993). *White queen psychology and other essays for alice.* Cambridge, MA: MIT Press.

6

Brain Studies: Trans Persons and Fish

Abstract Brain studies done in the last 25 years comparing Trans- and Cis-participants are reviewed. The research ranges over different brain areas and structures, with different investigations employing different measures. Strengths and weaknesses of the studies are highlighted. Three important issues emerge as the research progresses. Initially there was a search for brain-based etiologies for the Trans phenomenon. The next phase emphasized tracts and network differences, which in principle, can either bring about, or result from living a Trans experience. And, finally there are two issues more currently investigated: (1) the question of primary brain causes vs. epigenetic brain effects, and (2) the possibility of a new ontological trans phenotype. Studies of sex-changing fish (particularly one from Rhodes and colleagues [2019]) provide proof-of-concept.

Keywords Brain studies · Cause vs. effect · Epigenetic changes · Proof-of-concept: sex-changing fish · New ontological phenotype: Trans-men/Trans-women

L. A. W. Brakel, *Investigations into the Trans Self and Moore's Paradox*,
https://doi.org/10.1007/978-3-030-44645-1_6

1 Brief Introduction

Continuing with the biology portion of this project, several recent neuroscience brain studies with Trans persons (and Cis-controls) will be presented. Findings, along with appreciations, criticisms, and cautions will be offered. Indeed, significant findings do exist, but often results are variable across different studies, show group effects only, and of most importance, the findings are not able to dis-ambiguate brain causes vs. brain (epigenetic) effects of the Trans experience. This chapter will also include brain and behavioral research on sex-changing fish—particularly brain studies before, after, and during transitions—suggesting that this comparative species investigation can provide potentially useful data.

2 Humans

2.1 Search for Etiology

Physiologists and neuro-biologists, including Ai-Min Bao and Dick Swaab (2011, p. 214) have known for more than two decades that "…during the intrauterine period a testosterone surge masculinizes the fetal brain, whereas the absence of such a surge results in a feminine brain.[1] As sexual differentiation of the brain takes place at a much later stage in [fetal] development than sexual differentiation of the genitals, these two processes can be influenced independently of each other." Thus, more than two decades ago, the search began for brain variations in Trans persons that could be etiologically significant.

1. In 1995 an article titled "A sex difference in the human brain and its relation to transsexuality," Zhou et al. investigated a hypothalamic brain structure "essential for sexual behavior" (p. 68)—the bed nucleus of the stria terminalis or BSTc. Normally larger in males than

[1] Recently, this has been amended: A female brain is due to the absence of testosterone and the presence of other factors; only the combination results in a feminine brain. See McCarthy, M. and Arnold, A. (2011). "Reframing sexual differentiation of the brain." *Nature Neuroscience, 14*, 677–683.

in females, the researchers found (p. 29) that in their N of 6 male-to-female Trans persons, the BSTc was female-sized (in fact smaller than female control subjects, although this difference did not reach significance). Despite the small sample size, since the authors could discern no effect of blood levels of sex hormones, age, sexual orientation or age at which transition took place on these results, they felt their findings should stand, and suggested further investigations.

2. Then, in pursuing this result further, a team (including the above authors) lead by Kruijver et al. (2000) asked if the BSTc finding was due to genuine neuronal differences in the type of neuron most populating the BSTc—the multi-functional neurotransmission regulating "somatostatin-expressing neuron" (known also as Growth Hormone Inhibiting)—or did the finding owe to other unspecified factors. With 35 Cis-control subjects, the researchers established via post-mortem brain analysis that men had almost twice as many of the somatostatin-expressing neurons in their BSTc's as did women (p. 2034). Looking then at the same six male-to-female Trans persons from the 1995 experiment (since deceased), all showed a somatostatin neuron count in the female range, while the one female-to-male participant showed a count in the male range. (p. 2034, 2038). Again, finding no differences in neuron count on the basis of age of transition, size of brain, or sex hormones in the blood, the researchers concluded that these results could "…point to a neurobiological basis of gender identity disorder" (p. 2034). (This conclusion was reached rather too quickly, in my view, given the small sample size and other problems, with this and related studies. These problems will be taken up in a subsection just below).
3. A subsequent study by Swaab, with Garcia-Falgueras (2008) explored two subnuclei of the hypothalamic uncinate—interstitial nucleus of the anterior hypothalamus (INAH) 3 & 4. Analysis of post-mortem brain material on 35 Cis-control subjects revealed that the INAH-3 of males was almost twice (1.9 times) as big as that of females, and had more than two times (2.3) the number of neurons (p. 3132, 3138). Next the brain structures of 11 male-to-females Trans persons and 1 female-to-male Trans person were examined for INAH-3 size and

neuronal number. Regardless of age of transitioning, all eleven male-to-females were found in the female range on both measures, while the brain of the single female-to-male was in the female range in size and count. The authors then proposed that (1) These differences in the INAH-3 and the BSTc point to a "complex network that may structurally and functionally be related to gender identity" (p. 3132), and that (2) since fetal development of the brain is later than genital differentiation, these INAH-3 results and the earlier Trans persons' BSTc findings, when taken together "are hypothesized to be the basis of the [Trans] gender-identity…" (p. 3133). Three other findings from this study are of interest: The brains of the male-to-female Trans persons, were intermediate between the brains of male and female Cis-control subjects on measures of (a) volume (b) weight, and (c) neuronal density, significantly different from the male but not the female measures.

4. Bao and Swaab (2011), in the article cited above, summarize the findings from the above three studies. They also explain (again) that while the genitals are differentiated in the first trimester, the brain is "genderized" later, in the second and third trimesters, and can thereby be influenced independently of genital gender. The authors make the further claim that brain-gendering is also independent of post-natal cultural contexts, and is "irreversible" (2011, p. 215, 217).

Problems with BSTc and INAH-3 Studies

One glaring problem with all of these studies is the very small sample sizes. In two of the studies there is only a single female-to-male participant, and just six to eleven male-to-females. Another issue concerns the fact that while some of the subjects had undergone hormonal treatment, others had not. Although the researchers were well aware of this as a potential complication and attempted to control for these differences in various ways, it is unclear that the controls were sufficient. Another serious confound is that many of the male-to-female participants had been living as Trans women for many years. Thus, one would have to ask the following question regarding the brain analyses, both those done when the participants were living subjects, and later, on autopsy: Could at least some of the feminized brain effects owe to years of living as a woman, with constant, daily inputs of femaleness; including, but not limited to, perceptions of *Self* as a woman, both from self and others? Do the brain measures of male-to-female Trans persons in the 2008 study above, recorded as significantly different only from those Cis-male controls, but indeed intermediate between the Cis-males and Cis-females, support this? Given, the brain's plasticity, these seem plausible questions—questions we will explore further, as we continue our investigation of brain studies.

2.2 Network Studies

The neuroscience brain studies just presented, attempting to establish particular brain area changes as the most salient etiology for the Trans phenomenon, were inconclusive at best. As such, researchers began to explore a different idea with respect to differences between the brains of Trans and Cis persons: Perhaps the differences have to do with the neural networks underpinning awareness of *Self*, including both generalized self-awareness, and bodily self-awareness.

1. In a 2011 study Savic and Arver, noting the difficulties with post-mortem and small N samples, investigated several brain areas (but not the hypothalamus) in 48 Cis-controls—24 males, 24 females—and 24 male-to-female Trans persons, all untreated (with hormones). They found no differences between Cis-males and the male-to-female Trans persons in the following brain areas: Grey matter (GM) in the cerebellum and lingual gyrus; precentral gyrus; hippocampal volume; right cerebral hemisphere and thalamus volume (pp. 2525, 2528–2530). The researchers did find male-to-female Trans persons were significantly different from both male and female Cis-control participants in having (a) reduced volumes of the thalamus and the putamen[2];

[2]But, in an earlier f-MRI (functional-Magnetic Resonance Imaging) study, Luders et al. (2009, p. 906) found that that the GM of the putamen of untreated male-to-female Trans persons was larger than that of Cis-control males, within the average range of Cis-control female subjects.

and (b) elevated GM in the right insular and inferior frontal cortex, and in the areas covering the right angular gyrus (pp. 2525, 2528–2530).[3] Because these data from these brain findings show Trans participants to be not intermediate on these measures between the Cis-males and Cis-females, and in fact significantly different from both control subjects, the possibility of an ontologically separate Trans phenotype category (discussed briefly in Chapter 4, and in greater depth in Chapter 5's proper-function analysis) can gain some support. The authors of the study discuss another, no less important conclusion. Summarizing these results for the article's abstract, Savic and Arver (2011, p. 2525) state: "The present data do not support the notion that brains of Mt-F-TR (male-to-female Trans persons) are feminized. The observed changes in Mt-F-TR bring attention to the networks inferred in processing of body perception."

With respect to bodily perception networks, Savic and Arver (2011, p. 2530) do cite several relevant studies (some of which we will discuss just below) and, in addition offer the following (p. 2531): "One highly speculative thought is that the enlargement of the GM volume in the insular and inferior frontal cortex, and the superior temporal-angular gyrus could derive from a constant rumination about their own body…[with] [b]rain tissue enlargement in response…" Here, and with the bodily network investigations generally, the real possibility emerges that brain changes are not the cause of Trans-ness, but rather are the brain effects of the Trans experience. (Of course, interactions involving both cause and effect should not be ruled out.)

2. Pursuant to the views of Savic and Aver (2011) above, there are two kinds of network studies that do provide empirical support for the view that body and self-perception networks figure importantly in

This sort of variation in findings, even when the same imaging tool is used, contributes to feeling less than confident regarding the evaluation of brain differences between Trans individuals and Cis-controls. More about this below as we take up general problems with brain research involving Trans persons.

[3]The thalamus and insula are involved in sensory signaling (among other functions); the insula in particular is activated in self-recognition (See Craig 2009, p. 60). The putamen effects both bodily movement and category learning.

brain changes. And if that is the case, these changes could mark important differences seen in the brains of Trans vs. Cis persons.

I. The first type of network study shows that experiences, especially long term or intense ones, can affect brain changes. Although the studies cited do not involve the Trans experience, one should certainly regard years of living as a Trans person and/or contemplating such, as both quite intense and long in duration; in fact, more so than the experiences tested. There are two studies of this first type: Researchers Vestergaard-Poulsen et al. (2009) were able to demonstrate that there is increased gray matter in persons who have been long term meditators compared to controls; and Lewis et al. (2009) showed that there is plasticity in the adult brain, specifically in the visual cortex and frontal-parietal areas, consequent to repeated and intense visual perceptual learning tasks.

II. The second type of study is more directly relevant, as it concerns the brain connectivity involved with the *Self,* perceptions thereof, and a structural functional network, called the Default Mode Network, or DMN. Here are a few such studies in chronological order. In 2001 Gusnard et al. (p. 4259) found, using f-MRI: "The presence of self-referential *mental activity* [Gusnard's emphasis] appears to be associated with increases from the baseline in dorsal MPFC [Medial Prefrontal Cortex]." The authors continue that the activity in the MPFC region "…may contribute to the neural instantiation of the multi-faceted 'self' [and can] represent elements of the default state." This work is backed up by a 2006 meta-analysis of several such f-MRI studies run by Northoff et al. (p. 440) which concluded, "All studies revealed activation in the medial regions of our brains' cortex during self-related stimuli…[and] the activation…suggest[s] that self-referential processing is mediated by cortical midline structures." An f-MRI study investigating more specific brain networks was authored by Hodzic et al. (2009, p. 1264). They identified distinct cortical brain "…networks for body-detection (processing body related information), body-identification (processing of information relating to individual bodies), and self-identification (distinction of self from others)."[4]

The next three studies then, are important follow-ups, aimed at investigating whether Trans persons do show differences in these particular *Self* focused networks. If they do, and to the extent that they do, further questions obtain: are the demonstrated differences largely causes of Trans-ness or effects of Trans persons' *Self* experiences? Or is there a middle ground? Or is it different for different individuals?

3. Hypothesizing that Trans persons, in that they experience gender incongruity, would show brain effects reflective of this dissonance in their neural resting-state functional connectivity network (rsFC), Lin et al. (2014) compared f-MRIs of this network in 23 untreated Trans persons (11 male-to-females; 12 female-to-males) with 23 age-matched Cis-gendered control participants (11 males; 12 females). Using "graph theory-based network analysis," regional changes of the degree centrality of the rsFC network were recorded, with the degree of centrality indexing "the functional importance of a node in a neural network" (p. 1). As predicted the primary somatosensory cortex, and the superior parietal lobule did have a higher degree of centrality for the Trans persons (p. 2, 5). Another finding was that the connectivity between the right insula and the bilateral primary somatosensory cortex was negatively correlated with self-ness ratings of desired gender. In other words, the more participants rated themselves as being aligned with their Trans-gender, the less was the degree of centrality of the rsFC network, approximating that of the control group (pp. 6–7).
4. Taking the other position, namely, that brain network structural differences (rather than functional differences) in Trans persons might,

[4]Hodzic et al. (2009, p. 1264) found the following cortical networks: For body-detection: the cortical extrastriate body area (EBA) of both hemispheres, the Fusiform Body Area (FBA) of the right hemisphere, and areas of the inferior parietal lobe. Body-identification was found to involve the inferior frontal gyrus of both hemispheres, and a number of additional networks in the right hemisphere: the medial frontal gyrus, the cingulate gyrus, the central and post-central sulcus of the inferior parietal lobe, and the FBA. Distinguishing self from others involved differential activation of areas in the inferior parietal lobe, and inferior parietal sulcus of the right hemisphere, and in the posterior orbital gyrus and in the lateral occipital gyrus of the left hemisphere.

along with other brain changes, be important causal factors,[5] Hahn et al. (2015, p. 3527), "…investigated the structural connectome of 23 female-to-male…and 21 male-to-female patients before hormone therapy as compared with 25 female and 25 male healthy [Cis-]controls." The researchers found that both groups of Trans persons had decreased hemispheric connectivity ratios of subcortical/limbic areas compared with the controls (p. 3527, 3530). But the ratios were predicated differently for the two different Trans participants: Male-to-female subjects showed increased inter-hemispheric lobe connectivity weight; whereas for female-to-male subjects there was a decrease in intra-hemispheric lobe connectivity weight. The authors (p. 3530) conclude, first the obvious: That the two populations of Trans participants were clearly different from one another. Another conclusion, pertaining to the causal question is less clear-cut: The authors hold that the structural network findings of their studies, like the BSTc and INAH-3 brain structure findings they cite (see above), are reflective of "…structural differences in brain organization…already determined during pregnancy" (p. 3531).

5. In another study exploring functional connectivity, Feusner et al. (2017) suggest more clearly that brain network changes are effects of living a Trans life experience rather than causes of Trans-genderedness. Also, the authors again allow the possibility of a new ontological phenotype category, finding that Trans-ness—at least for those Trans persons with Gender Dysphoria—is not something midway between male-ness and female-ness. They state that their findings "…do not support the hypothesis that sexual differentiation of the brain in individuals with GD [Gender Dysphoria] is in the opposite direction as their sex assigned at birth; [specifically] the weaker…connectivity within the DMN [functioning in self-referential and own body perception] was detected in comparison with both male and female controls, and there were no differences between the two control groups" (p. 972). The experiment involved 27 untreated female-to-male Trans participants and 54 control participants—27 Cis-males, 27 Cis-females. Not only did the Trans participants show

[5]The authors are careful to point out that their study cannot definitively establish this.

decreased connectivity within structures (anterior and posterior cingulate and precuneus) of the DMN (p. 969, 971), but also visual networks in occipital and temporal regions involved with own body perception showed dysconnectivity (p. 970). Then, finally, with a technique called "the body-morph test"[6] participants viewed photographs of their bodies morphed to appear more masculine or more feminine, and rated each in terms of self-identification. The more that female-to-male Trans participants self-rated their bodies as morphed toward male-ness, the stronger was the connectivity (specifically of the anterior cingulate) within their DMN (p. 968, 970). In other words, as these subjects viewed self-images more in line with their Trans-gender, the less self-incongruence they seemed to experience.

Problems with the Network Studies

Eliminating one of the serious confounds in some earlier investigations, all of the participants in the three network follow-up studies were untreated, both in terms of hormones and surgeries. Still there are two major problems involving the subjects in these network studies, and they are intertwined. Most of the participants are not merely Trans persons, they are either explicitly diagnosed with Gender Dysphoria, or selected (recruited) on a similar basis, sometimes even referred to as "patients." Not only is Gender Dysphoria, as a diagnosis undergoing possible changes (See Chapter 1, Footnote 5), but also, one must wonder if the selection of such participants is actually representative of Trans persons. In other words, are the investigations skewed in a particular "pathological" direction. This leads to the related problem: Are the findings specific to different brain phenomena found in Trans persons, or do they reflect secondary problems that only some Trans persons have in being Trans persons?

[6]The Feusner group's "body morph test" would also be very useful in some of the empirical Experimental Philosophy thought experiments proposed in Chapter 7, where classic *Self* Thought Experiments are adapted to explore issues central to Trans persons.

A further, not unrelated, and perhaps even more troubling issue pertains to questions of cause or effect. Is Trans-ness caused by brain differences owing to differential brain and gonad fetal development? Or are the brain changes effects of one or more of the following experiences: (a) living as a Trans person; (b) living with the knowledge that wants to or intends to transition; (c) experiencing a major incongruity within one's *Self*; (d) feelings involved with this specific incongruity? Or is there a mixture—with some brain changes causing feelings of Trans-ness, which when experienced cause different secondary brain changes. These matters could, in principle, be disambiguated in future experiments, but not without considerable difficulty, including the problem of recruiting representative subject-participants in sufficient numbers.

These network studies also raise additional central questions. Related to the cause or effect problem, there is the question of timing: If the brains of Trans persons are feminized in the case of male-to-females, or masculinized for female-to-males, does this take place as a result of pre-natal development; or is this brain-changing an epigenetic post-natal phenomenon? Then an ontological category question: Is the feminization of male-to-female brains (and a degree of de-masculinizing), and the masculinization of female-to-male brains (and some de-feminizing) merely a quantitative type brain response, approaching the new gender from the natal one, and ending up midway; or do the particulars of these processes represent a new ontological phenotype category—consisting of Trans-men and Trans-women. The next group of brain investigations presented attempt to address these questions.

2.3 Two Important Issues: New Ontological Phenotype Category? and Cause or Effect?

1. A study of cortical thickness by Zubiaurre-Elorza et al. (2013) although explicitly agnostic, presents evidence that could weigh in favor of the new ontological phenotype hypothesis. Contrasting MRI findings that (a) Female Cis-controls (n = 23) have thicker frontal and parietal cortical regions than do Male Cis-controls (n = 29), and (b) that Male Cis-controls have greater right putamen volume

than the Female Cis-controls, the researchers investigated the MRIs of 24 female-to-male and 18 male-to-female Trans participants, all untreated. They found that the female-to-male participants were similar to the female controls in overall cortical thickness, but had right putamen values larger than the female controls, in fact in the male range. The authors deem this "subcortical gray matter masculinization." The male-to-female participants showed a different pattern in that on overall cortical thickness they were in the female range, a change considered "cortical thickness feminization" (p. 2855, 2859). All the significant differences were in the right hemisphere. The researchers conclude (p. 2860): "We would suggest that transsexuals do not show a simple masculinization (FtM [female-to-male]) or feminization (MtF [male-to-female]) of their brains—rather, they present a complex picture in their process of sexual differentiation depending on the brain region studied and the kind of measurements taken."

Regarding the cause vs. effect question, the researchers lean toward the view that it is pre-natal brain differences that cause Trans-sexuality (p. 2860): "In both types of transsexuals, the pattern of differences with respect to their biological [natal] sex is located in regions of the right hemisphere. This might suggest a hemispheric asymmetry in the effect of androgens during the development of the brain in transsexuals."

2. Data that suggest the possibility for a distinct ontological phenotype category for Trans persons (although the authors do not use these terms) were found in a study in which Simon et al. (2013) analyzed MRI data with "voxel based morphology" to determine whether and where significant differences in grey matter structure(s) obtained between 17 Trans persons (10 male-to-females and 7 female-to-males), all untreated, but with the diagnosis of Gender Identity Disorder (GID)[7] and 18 age matched Cis-controls (11 females, and 7 males). The researchers showed (p. 8) that the Trans participants as a group (including both male-to-females, and female-to-males) did significantly differ from the control participants as a group (both Cis-males and Cis-females) in the left angular gyrus, and the left inferior

[7] See again Chapter 1, Footnote 5.

parietal lobe (p. 1, 4). Also, significant differences between Trans participants and controls were found in both the anterior and posterior lobes of the left cerebellum, a finding both robust, and to that point undocumented in the literature (p. 8).[8] Further results were obtained in several regions in which the grey matter volume of the Trans participants matched with those sharing their gender identity, not those of their natal sex. These include: left and right precentral gyrii, the left postcentral gyrus, the left posterior cingulate, precuneus and calcarinus, the right cuneus, the right fusiform, lingual, middle, and inferior occipital, and inferior temporal gyri.

Discussing their findings, the authors state (p. 8): "The regions found in our study are mainly involved in neural networks playing a role in body perception, including memory retrieval, self-awareness, visual processing, body and face recognition and sensorimotor functions. Findings from the studies of Savic and Arver [study done in 2011; cited above] and Luders et al. [study done 2009; see footnote 2 in this chapter] also indicated that these functional networks are affected. We must note, however, that it is still unclear how the observed structural differences might translate into functional differences." The Simon et al. (2013) article continues, with careful but astute comments on the cause vs. effect dilemma (p. 8): "Furthermore, it is also unclear how the observed differences could be interpreted in terms of a causal chain: as parts of the neurobiological background or perhaps as consequences of transsexualism."[9]

In order to highlight the difficulty in discerning whether or not a distinct ontological phenotype category for Trans persons exits, let

---

[8]A later study partly corroborated the cerebellar findings. Contrasting grey matter (GM) volumes in various brain structures between Trans adolescents (55 female-to-males; 38 male-to-females; all with GD) and 44 cis-boys and 52 cis-girls (without GD), Hoekzema et al. (2015, pp. 59–60) found "…less GM volume in the right cerebellum…in female-to-males in comparison to girls without GD, while male-to-females had less volume in the bilateral cerebellum …than natal boys." Note that the cerebellar differences in the Hoekzema (2015) study are bilateral and left, whereas in the Simon (2013) study, the findings are just in the left cerebellum.

[9]This issue is once again made more complicated owing to the same problem raised for the network studies—the Trans participants have a "diagnosis," hence confounding disorder/pathology with trans-ness.

me illustrate the importance of researchers' interpretative views, as we consider the next two studies.

3. Investigating white matter tracts using diffusion weighted MRI analysis in 23 female-to-male Trans participants, 21 male-to-female Trans participants, 23 female Cis-controls, and 22 male Cis-controls, Kranz et al. (2014) found that there were significant differences in the measure of interest—mean diffusivity—in almost all white matter tracts. "FCs [Female-controls] had highest mean diffusivities, followed by FtM [female-to-male] transsexuals with lower values, MtF [male-to-female] transsexuals with further reduced values, and MCs [male-controls] with lowest values" (p. 15466). In the discussion section, the authors report (p. 15471): "The results of this study show that white matter microstructure in FtM and MtF transsexuals falls halfway between that of FCs and MCs…" The rank order of mean diffusivity being: Female-Cis controls > Female-to-males > Male-to-females > Male-Cis controls (p. 15471). Although the authors do not explicitly state this, they interpret this scalar ordering as aligned both with the view that pre-natal brain/gonad disconnection causes the experienced gender incongruity (p. 15472, 15474); and, moreover, that this scalarness weighs against any new distinct ontological phenotype. Instead, for these authors, what seems to be the case is an unfolding linear and quantitative (rather than qualitative) process, in which the brains of female-to-male Trans persons undergo some combination of masculinization and de-feminization, while the brains of male-to-female Trans persons likewise are in part de-masculinized and also feminized (p. 15474).
4. A view very different from that of the Kranz et al. (2014) study just above, comes from Guillamon et al. (2016) in their review article on the status of research on the Trans-sexual brain. The authors assert (p. 1627): "Untreated…MtFs [male-to-females] and FtMs [female-to-males] show a complex picture for the expression of sex differences in their brains…Contrary to some popular ideas, the MtF brain is not completely feminized but presents a mixture of masculine, feminine, and demasculinized traits…Moreover, the brain of …FtMs is

not uniformly masculinized but presents a mixture of feminine, defeminized, and masculized morphological traits…Thus, the morphology of the brain of …MtFs and FtMs strongly suggest that each one has its own phenotype, and that the phenotype is different from those of …[cis] males and females."

The view presented by Guillamon et al. (2016) allows for the possibility that the social environment—for example lives lived experiencing Trans-ness—can be, via epigenetic influences on gene expressions within the brain, an important cause of the structural/function brain changes documented in Trans individuals. After all, brain plasticity is considerable in humans.

And since brain plasticity is perhaps even greater in some fish species, we will in the next section, turn to a study with a sex-changing fish species. This will be undertaken in order to advance a proof-of-concept type of analogy in that fish researchers can demonstrate that social environmental cues, perceived by fish individuals, lead to epigenetic influences on the gene expression within that individual's brain, which in turn lead to body morphological and finally gonadal sex change. But before we get to fish, we need to take up problems with the research just outlined in the "Two Issues" section on human brain studies.

Problems with the Research Related to the "Two Issues"

With regard to the question of distinct Trans ontological phenotype categories—Trans-man/Trans-woman—the difficulties with participant selection discussed above continues and is even more problematic. Namely, since all of the Trans participants in the studies reported in this subsection carry a diagnosis—Gender Dysphoria (GD) or Gender Identity Disorder (GID)—and as these diagnoses are focused (as the names indicate) on dysphoria and disorder, any result confounds Transness with at least some level of psychiatric pathology or problem. There is a potential solution: Recruit two additional contrast groups of participants: (a) Trans persons who are relatively unconflicted, and (b) Cis persons, age and socio-economically matched, who suffer from distress—both of a gender related type[10] and dysphoria not particular to gender.

On the other matter—that of cause vs. effect—perhaps investigating participants at different stages of their Trans experience would prove helpful. For instance, if post-natal experiences in living life as a Trans person were central in causing epigenetic brain change, Trans persons experiencing their Trans-ness longer (beginning in childhood or early adolescence rather than in adulthood) might show more pronounced brain change effects.

3 Fish

3.1 Anemonefish—Socially Driven Epigenetic Brain Changes—A Proof-of-Concept Account

Amenonefish (*Amphiprion ocellaris*) is a species whose adult members, according to Francis (1992, p. 2), "…live largely in pairs with associated subadults. Here the larger dominant member of the pair is female; if she is removed, her mate will change sex if the territory is not taken over by a larger male…" In Justin Rhodes lab, Dodd and colleagues (2019, pp. 65–76) experimentally induced this situation as follows (p. 66): Reproducing pairs of amenonefish, 17 males and 17 females were separated from their mates. Then the males were randomly paired together. Observations and planned measurements were taken after 3-week, 6-month, 1-year, and 3-year intervals. However, after just the first few days, in every pair but one, a dominant fish clearly emerged, with greater size always determinative of dominance (p. 68). Moreover, in each of the 16 pairs the bigger dominant fish grew faster (p. 69). Also, and rather strikingly, in terms of basic morphology, the newly dominant fish appeared to be unremarkably female, this concurrent with its overt dominance.

[10]Without a doubt, Cis persons can experience gender related dysphoria without any feelings of gender incongruity. For example, males may feel undue pressure to perform sexually, or to match up with certain masculine stereotypes. Females, in parallel, may feel distress over "typical" feminine issues, such as concerns about beauty, or maternal instincts.

The brain responses to this social situation were also reported. In this species of fish there is sexual dimorphism in the preoptic area (POA) of the hypothalamus, in which females have a greater number of medium sized cells.[11] And here the brain results were quite remarkable, if not as rapid: Feminization, indicated by an increased number of medium cells in the anterior POA, took place incrementally.[12] The 3-week comparison between the subordinate males and the newly dominant members showed no significant difference in the number of POA medium cells. But, by the 6-month and 1-year tests, there were significant differences between the subordinate males and the dominant bigger fish; with the latter individuals now at female level POA medium cell counts (p. 72), while the subordinate fish remained at male levels (p. 71). The researchers state that these results (p. 72): "…establish that the number of cells in the anterior POA increase monotonically as a male transforms into a female over a period of 6 months to 1 year before stabilizing at female levels in *A. ocellaris.*"

Gonadal changes to complete feminization—which, as defined by the experimenters is (p. 69) "…the presence of at least some vitellogenic oocytes in the gonad of and individual which was once male"—only took place in three of the 17 pairs, sometimes as long as years after feminization of the POA (p. 70). Regarding this finding, the researchers' comment (p. 69), "…that gonadal sex change in *A. ocellaris* occurs on a variable time-scale and requires some unknown stimulus to occur which is either individual or context specific." This idea, that specific somatic instantiations (here gonadal) consequent to epigenetic brain changes are quite variable, is one that is quite likely applicable to other species, including humans.

---

[11]The POA is an area regarded as homologous to the structures of human hypothalamic nuclei and sub-nuclei explored above in this chapter "Searching for an etiology"—Studies 1–4. The Rhodes research group states in Dodd et al. (p. 74) that the POA "…contains the parvocellular nucleus in fishes which is considered homologous to the sexually dimorphic regions of the POA in mammals, birds, reptiles, and amphibians."

[12]The medium cells in the POA are thought to be dopamine neurons or other neurons expressing Esr1 (an estrogen receptor) projecting onto and thereby regulating GnRH (gonadatrophin releasing hormone) neurons. Although still being researched, if this is found to be among the functions of the POA medium cells, Dodd et al. state that the sexual dimorphism demonstrated could translate readily to mammals (p. 23).

Perhaps it is surprising, given the nature of this project as a whole, that the gonadal status of these fish, while interesting, is of less import than the proof-of-concept analog for environmental, socially perceived epigenetic brain changes that this study amply provides. The researchers in the Rhodes' experiment, changed the social environment of the fish, splitting up adult reproductive couples and then pairing sets of males together randomly. And the fish, upon clearly perceiving the social change, responded first with very rapid body morphology in the direction of male-to-female sex-change. This suggests (but does not prove) that the direction of the causal chain goes from (1) Social Situation Change and Perception Thereof→ (2) Body Change→ (3) Brain Change. This is not at all the brain-change first causal etiology forwarded by some of the human brain researchers discussed above.[13]

Finally, so as not to have our fish data focus solely on male-to-female sex-changing fish, consider a species in which the sex-change is from female-to-male, the bluehead wrasse. As described by Godwin (2010, p. 208), social change for these fish, namely the removal of the dominant male in a group, begins a process for the largest female to transition to male-ness. This socially salient environmental change leads to significant changes in the POA, specifically "…rapid increases in AVT mRNA…." (p. 208). Importantly, note that AVT (arginine vasotocin) and AVT neurons in these wrasses "appear especially important in promoting male-typical behavior such as courting and aggression."—this according to fish researchers, Todd et al. (2016, p. 231).

Evaluating the Proof-of-Concept Account for Epigenetic Change

Speaking generally now, and not just of fish, it is well known that epigenetic regulation of gene expression is critical during development. Todd et al. (2016, p. 235) add that epigenetic modification is also "…a mechanism through which environmental cues can be translated into plastic phenotypic responses." These same authors, now focusing on fish, specifically fish of the sex-changing species, maintain that (p. 235):

[13]An important caveat: fish findings do not necessarily generalize to humans.

"...environmentally-induced sex reversal likely involves sex-specific epigenetic reprogramming of the genome and may include pervasive DNA methylation to silence the opposing sexual pathway."

Constructing an analogy between sex-changing fish and Trans humans, consider that constant environmental and experientially triggered epigenetic brain changes seem possible, perhaps even likely in persons (a) feeling/knowing that their gender is wrong; (b) contemplating transitioning; and/or (c) living for years as Trans persons. By the same token, the life stories of sex-changing fish, do not preclude another analogy, paradoxically leading to a conclusion in the opposite direction. In both the male-to- female fish (protandry), and the female-to-male fish (protogyny), the sequence established by research is as follows: (1) Social Environmental Change→ (2) Perception thereof→ (3) Changes in Body Morphology, then Changes in Brain Morphology→ (4) Gonadal Change. But note, although indeed the brain alterations are initiated by the registration of a social change and do occur after body morphology changes, the brain changes do precede gonadal changes. This clear separation of brain and gonadal sex, and the causal chain direction thereof, could be taken as a proof-of-concept analogy for those researchers favoring brain differences as the central causal etiology for the experience of Trans-ness.

And so, as with many issues in this project, we must conclude not to conclude. Instead, we should explore individual differences, both among people and fish; and more generally, try to promote further research by welcoming multiple hypotheses, even those originating from divergent points of views.

* * *

The chapter on Brain Research (This chapter), has not only provided much useful data and many interesting findings, but also points the way toward investigating many unanswered questions and reveals more vexing puzzles. In looking for an etiology for Trans phenomena we discover a cause and effect loop—are pre-natally derived brain structures different for Trans persons, and are these differences causative? Or are the sorts of structural/functional changes, for example in several of the networks and tracts, particularly those related to *Self*—perception, recognition—effects, rather than causes, of a Trans- gender experience? Perhaps there is a different blend for individual Trans persons. But these individual effects cannot be discerned from much (if not all) of the human research cited above, because although these neuroscience studies were well conducted, they investigate different brain structures and networks, with different measures, and compile only group effects—at best differences between each of the two Trans groups (male-to-female and female-to-male) compared with each of the two control groups (Cis-males and Cis-females).

Another dilemma concerns whether or not Trans-ness in humans (Trans-men and Trans-women) can represent a distinct new ontological phenotype category—this within the standard binary gender system. While some of gender-theorists (including some discussed in Chapter 4) might endorse this idea, and some of the human brain research reported upon can be viewed as supporting this concept, there are no clear answers. Perhaps when the Trans subjects are midway between the Cis-controls on scalar quantitative measures, one would say no new category; but when the results from the Trans participants are uniquely different from both male and female Cis-control subjects, one might want to establish a new ontological phenotype. The next chapter, Chapter 7, consists of a series of empirical Experimental Philosophy (X-Phi) Thought Experiments. Adapted from classical Thought Experiments on the *Self,* they may afford some further insight into the question of a new Trans phenotype category as well as lead to surprisingly useful diagnostic information.

Bibliography

Bao, A.-M., & Swaab, D. (2011). Sexual differentiation of the human brain: Relation to gender identity, sexual orientation, and neuropsychiatric disorders. *Frontiers in Neuroendocrinology, 32,* 214–226.

Craig, A. (2009). How do you feel—now? The anterior insula and human awareness. *Nature Reviews/Neuroscience, 10,* 58–70.

Dodd, L., Nowak, E., Lange, D., Parker, C., DeAngelis, R., & Rhodes, J. (2019). Active feminization of the preoptic area occurs independently of the gonads in *Amphiprion ocellaris. Hormones and Behavior*, 112, 65–76.

Feusner, J., Lidstrom, A., Moody, T., Dhejne, C., Bookheimer, S., & Savic, I. (2017). Intrinsic network connectivity and own body perception in gender dysphoria. *Brain Imaging and Behavior, 11,* 964–976.

Francis, R. (1992). Sexual lability in teleosts: Developmental factors. *The Quarterly Review of Biology, 6,* 1–18.

Garcia-Falgueras, A., & Swaab, D. (2008). A sex difference in the hypothalamic uncinate nucleus: Relationship to gender identity. *Brain, 131,* 3132–3146.

Godwin, J. (2010). Neuroendocrinology of sexual plasticity in teleost fishes. *Frontiers in Neuroendocrinology, 31,* 203–216.

Guillamon, A., Junque, C., & Gomez-Gil, E. (2016). A review of the status of brain structure research in transsexualism. *Archives of Sexual Behavior,* 45, 1615–1648.

Gusnard, D., Akbudak, E., Shulman, G., & Raichle, M. (2001). Medial prefrontal cortex and self- referential mental activity: Relation to a default mode of brain function. *Proceedings of the National Academy of Sciences, 98,* 4259–4264.

Hahn, A., Kranz, G., Kublbock, M., Kaufmann, U., Ganger, S., Hummer, A., et al. (2015). Structural connectivity networks of transgender people. *Cerebral Cortex, 25,* 3527–3534.

Hodzic, A., Kaas, A., Muckli, L., Stirn, A., & Singer, W. (2009). Distinct cortical networks for the detection and identification of the human body. *NeuroImage, 45,* 1264–1271.

Hoekzema, E., Schagen, S., Kreukels, B., Veltman, D., Cohen-Kettenis, P., Delemarre-van de Waal, H., & Bakker, J. (2015). Regional volumes and spatial volumetric distribution of gray matter in the gender dysphoric brain. *Psychoneuroendocrinology, 55,* 59–71.

Kranz, G., Hahn, A., Kaufmann, U., Kublbock, M., Hummer, A., Ganger, S., et al. (2014). White matter microstructure in transsexuals and controls investigated by diffusion tensor imaging. *The Journal of Neuroscience, 34,* 15466–15475.

Kruijver, F., Zhou, J.-N., Pool, C., Hofman, M., Gooren, L., & Swaab, D. (2000). Male-to-female transsexuals have female neuron numbers in a limbic nucleus. *The Journal of Clinical Endocrinology and Metabolism, 85,* 2034–2041.

Lewis, C., Baldassarre, A., Committeri, G., Romani, G., & Corbetta, M. (2009). Learning sculpts the spontaneous activity of the resting brain. *Proceedings of the National Academy of Sciences, 106*, 17558–17563.

Lin, C., Ku, H., Chao, H., Tu, P., Li, C., Cheng, C., et al. (2014). Neural network of body representation differs between transsexuals and cissexuals. *PLoS ONE,* 9, e85914, 1–10. https://doi.org/10.1371/journal.pone.0085914.

Luders, E., Sanchez, F., Gaser, C., Toga, A., Narr, K., Hamilton, L., & Vilain, E. (2009). Regional gray matter variation in male-to-female transsexualism. *NeuroImage, 46*, 904–907.

Northoff, G., Heinzel, A., deGreck, M., Bermpohl, F., Dobrowolny, H., & Panksepp, J. (2006). Self-referential processing in our brains—A meta-analysis of imaging studies on the self. *NeuroImage, 31,* 440–457.

Savic, I., & Arver, S. (2011). Sex dimorphism of the brain in male-to-female transsexuals. *Cerebral Cortex, 21,* 2525–2533.

Simon, L., Kozak, L., Simon, V., Czobor, P., Unoka, Z., Szabo, A., & Csukly, G. (2013). Regional grey matter structure differences between transsexuals and healthy controls—A voxel based morphology study. *PLoS ONE, 8,* e83947. https://doi.org/10.1371/journal.pone.0083947.

Todd, E., Liu, H., Muncaster, S., & Gemmell, N. (2016). Bending genders: The biology of natural sex change in fish. *Sexual Development, 10,* 223–241.

Vestergaard-Poulsen, P., van Beek, M., Skewes, J., Bjarkam, C., Stubberup, M., Bertelsen, J., & Roepstorff, A. (2009). Long-term meditation is associated with increased gray matter density in the brain stem. *NeuroReport,* 20, 170–174.

Zhou, J.-N., Hofman, M., Gooren, L., & Swaab, D. (1995). A sex difference in the human brain and its relation to transsexuality. *Nature, 378,* 68–70.

Zubiaurre-Elorza, L., Junque, C., Gomez-Gil, E., Segovia, S., Carrillo, B., Rametti, G., et al. (2013). Cortical thickness in untreated transsexuals. *Cerebral Cortex, 23,* 2855–2862.

Part IV

Future Investigations

7

Experimental Philosophy (X-Phi) Studies

Abstract This chapter makes good on the promise (from Chapter 3) to adapt for empirical Experimental Philosophy studies a number of Classic Thought Experiments on the nature of *Self.* The Body-Brain Exchange (B.Williams); Closest (now Real) Continuer (Nozick); and Tele-transporting with Percentage change (Parfit) have all been re-formulated to explore Trans vs. Cis differences, and more importantly, suggest diagnostic and prognostic implications for different groups of Trans persons. Also, the experiments are devised to gain evidence for (or against) the new ontological phenotype—Trans-men/Trans-women. About this hypothesis a final and simple Moore's Paradox study could prove revealing.

Keywords Experimental philosophy empirical studies · X-Phi adaptations of classical *self* thought experiments · Diagnostic and treatment implications · New ontological phenotype: Trans-men/Trans-women · Moore's Paradox re-visited · Future directions

L. A. W. Brakel, *Investigations into the Trans Self and Moore's Paradox*,
https://doi.org/10.1007/978-3-030-44645-1_7

1 Preliminary Considerations

Experimental Philosophy (X-Phi) is a relatively new branch of philosophy that marshals empirical methods, often toward gaining data-based answers to thought experiment questions. The process goes something like this: After researchers have registered their hypotheses, the thought experiment questions are shaped by experimental design criteria, including the need for control conditions and constraints to ensure sufficient balance among potential responses. When the design criteria are met, research subjects are recruited, the experiment run, and then the accrued data analyzed using statistical measures that have been shown to be reliable and valid. Again, this is all in the service of gaining some traction on, if not actually answering, hard theoretical questions. The nature of *Self* is such a hard question—and as we've seen through the various chapters of this work, the three foundational philosophical questions arising with the Trans experience have yet to be sufficiently answered. It is my hope then, that in this X-Phi chapter, the Empirical Thought Experiments I have devised—derived from classical *Self* Thought Experiments, and adapted in order to compare Trans and Cis individuals—might open avenues to further knowledge, not only about Trans persons, but about the *Self* itself. Thus, through these studies, if better theories of the *Self* can be advanced, this X-Phi chapter might turn out to be the most important contribution of this monograph.

It is important to state at the outset, that for these Empirical Thought Experiment studies, as for this entire project, we will be considering only Trans participants who live within the existing mainstream binary gender system, endorsing it, but with the knowledge and deeply held conviction that they are the "wrong" gender. This limiting case allows for a clearer, sharper investigation, and it is only on these practical grounds that Trans persons who identify as fluid-gendered, mixed-gendered, non-gendered, or unspecified third-gendered are not part of the anticipated participant pool in any of the proposed Experimental Philosophy Empirical Experiments.

Concerning the particular studies to be presented in this chapter: (1) They all consist of empirical responses and analyses to particular Thought Experiments, where the Thought Experiments are adapted to address issues of the *Self* highlighted by the Trans experience; (2) The empirically treated Thought Experiments are derived from classical philosophical Thought Experiments on the nature of *Self*; and (3) The original classical Thought Experiments have been discussed previously in this project, in Chapter 3, in the subsection on the *Self* one cares about when one cares about one's survival.

As a final preliminary comment, it should be noted that all of the studies in this chapter are proposed with the goal of progress (or at least the possibility of progress) toward:

1. Providing potential evidential support for (or against) the new ontological phenotypes—Trans-men and Trans-women. These are phenotypes additional to the regular categories of Men and Women,[1] but still within the mainstream binary gender system, which is the limiting case of this project.
2. Diagnostic implications for Trans persons: First, the experiments might help to distinguish different groups of Trans persons—(a) those who feel and know that their gender incongruence is located mostly in the brain/mind; (b) those who feel and know that the non-fit is largely body centered; and (c) those who, despite a feeling of gender disconnection, might be disinclined to pursue a Trans experience into early or mid-adulthood.[2] Second, across all three of these groups, for every participant as part of the pre-screening, a simple baseline instrument measuring general happiness/unhappiness, as well as gender content/discontent, should be given.[3] This would go some distance in disambiguating Trans-ness, as an experienced phenomenon,

---

[1]If the proper-function argument of Chapter 5 goes through, the more inclusive categories—MEN (to include Trans-men and Cis-men) and WOMEN (to include Trans-women and Cis-women)—would indeed be considered the regular (Normal) proper-functioning binary system categories.

[2]Hahn et al. (2015, p. 3531) reports that 75% of Trans identified children and adolescents who present with a diagnosis of Gender Dysphoria do not pursue the Trans experience into adulthood. But see again Chapter 1, footnote 5 outlining the research confound regarding any psychiatric pathology diagnosis, including Gender Dysphoria.

[3]There are many such instruments in current concerning general happiness/unhappiness. It is likely that a comparable measure designed specifically to assess gender contentment/discontent could be readily constructed.

from some level of Gender Dysphoria which may or may not be occasioned.

2 The Experiments

2.1 The Body/Brain Exchange—Experiments One and Two

Overview

The Body/Brain Exchange—Experimental Philosophy Version, is an empirical study. It is adapted from the classic *Self* Body/Brain Exchange Thought Experiment attributed to Bernard Williams (1970/1973, pp. 51–52) and Sidney Shoemaker (1963, p. 23); commented upon by Thomas Nagel (1986, p. 42); and discussed at some length in the current work (Chapter 3). The X-Phi version, which follows immediately below, is designed to explore whether or not there exists two separate groups of Trans persons—those who view the disconnect as located mostly in the body, vs. those who experience the incongruence to be largely brain/mind based. Relatedly, the experiment aims to investigate the *Self* views of Cis persons, both as control subjects, and to ascertain as a matter of interest what percentage identify their deepest *Self* as brain/mind-based vs body-based.

The experiment—actually two experiments, first one for children and adolescents, and the other just alike, except designed for adult participants—would each be run in two parts. Part One, for the Cis-controls, more closely resembles the original classic Thought Experiment than the version adapted for Trans participants, but in both Trans and Cis versions there is one major change from the original classic in that the prospect of future pain is totally omitted.

Here then is story told to the Part One Cis participants: "Person-A and Person-B will be required to exchange their bodies and brains. (Brains will of course include all mental contents.) Although Person-A and Person-B are similar in body and brain, much as twins might be, they are distinct individuals. The good news is that surgical science has progressed remarkably such that these procedures can be performed without pain or any untoward side effects. Now, suppose you start as Person-A with A-Body and A-Brain. The surgery must take place, but you can choose to keep A-Brain and have it put in B-Body; OR keep A-Body and have B-Brain installed. Which do you prefer? Which is more your real-*Self*, your me?"

Part Two, has been altered a bit for young Trans persons, and there would be a male-to-female story and a female-to-male one. Starting with our emblematic Trans person G→B, the researcher would say: "You must participate in the following Brain/Body Exchange experiment. The good news is that surgical science has progressed remarkably such that body/brain exchanges can be done without any pain or complications. So, here is the choice for you, GB: You can either keep your current Brain and have it transferred into a Body that is a lot like yours except that it is a male twin/clone; OR you can keep your current Body, and have installed into it a Brain twinned/cloned from yours, except that this brain feels itself/ knows itself to be female. Which do you prefer? Which is more your real-self, your me?"

In both Parts One and Two of the study, just after the experimental set-up story has been understood, subjects will be asked to describe their feelings, and be given both a basic "mood thermometer" measure[4] and an analogous instrument to assess subjective gender content/discontent.

The data analysis would consist of several paired comparisons for the responses—Body or Brain. The comparisons include: (1) Cis participants with Trans participants; (2) male-to-female Trans persons with female-to-male Trans persons; (3) male-to-female Trans persons with Cis-males; (4) male-to-female Trans persons with Cis-females; (5) female-to-male Trans persons with Cis-females; (6) female-to-male Trans persons with Cis-males; and finally (7) Cis-males with Cis-females. Also, any differences in mood and affect among these groups would be of interest. And, here is one hypothesis on this matter: While most Cis persons will find this experiment awful or at least unpleasant, there might be an interesting

[4]General measures are already available. See for example, Zavala-Rojas, D. (2014) "Thermometer scale (feeling thermometer)." In A. Michalos (ed.), *Encyclopedia of quality of life and well-being research*. Dordrecht: Springer.

contrast with at least some Trans individuals—especially those registering much discontent on the gender scale. Namely, there might be some Trans persons who could imagine allowing, tolerating, and enduring such an event not with dread, but rather with a feeling of hopefulness at the prospect of painless transplant of either body or brain, to effect gender congruence.

These experiments could prove of some diagnostic significance with regard to distinguishing two groups of Trans persons—those who perceive the locus of incongruity in the brain(/mind), and those who find the body to be "wrong." Different subsequent interventions could be deemed appropriate in each case—ranging from physiologic/physical transitioning, at one extreme, with psychotherapy at the other. Moreover, participants' affectual response might also provide some diagnostic help in predicting which persons could benefit most rapidly, and which might need further therapeutic discussions. Note that running this experiment for children and adolescents, given its potential for diagnostic implications, could be particularly beneficial.

Having set out the basic overview of the experiment, what follows is the presentation of it in detailed empirical format. Included are also some ideas for future directions.

The Experiment—Presented in Empirical Format

Experimental Philosophy Experiment One: Body/Brain Exchange for Children and Adolescents

PART ONE—Cis-Gendered

Design

Participants:
100 Cis-Gendered Control Participants, 50 Males; 50 Females; all children and adolescents under 18 years old, matched for age, socio-economic, and psychological factors to the Trans-Gendered Participants of Part Two.

Set Up:
Participants are told a version of the classic Body/Brain Transfer thought experiment:

"Person-A and Person-B will be required to exchange their bodies and brains. Suppose that surgical science has developed remarkably such that without any pain or complications brains and bodies can be switched. Now suppose that you start the experiment as Person-A. The surgery must take place, but you can choose either to keep Brain-A and get it transplanted into Body-B; or keep Body-A and get Brain-B put inside. (Participants should be assured that Person-A and Person-B are similar to one another in both body and brain the way "identical" twins would be, but they are distinct individuals.)

Procedure:
Immediately after the "set up" story has been sufficiently understood, the experimenter will ask every participant to describe their feelings (in as much detail as possible) as the reality of the experiment is imagined. A simple mood thermometer measure will be given. Then each participant will be asked: "Which is more your real-*Self*, your me? Put another way: Which remains as you—the original body or the original brain—and which is transplanted?"

Analysis

1. Look at group statistics—which is more real-self, me? Does body or brain remain?
2. Compare Males with Females
3. Assess affect and mood
4. Compare Males with Females on affect assessment and mood measure

PART TWO—Trans-Gendered

Design

Participants:
100 Trans children and adolescents—ages under 18; and all untreated medically with respect to gender. 50 Male-to-Females; 50 Female-to-Males

Set Up:
Participants are told one of two versions of the below story. One for male-to-female Trans persons and one for female-to-male Trans Persons. Below is the female-to-male version, given GB as our emblematic person.

"You must participate in the following Brain/Body Exchange experiment. The good news is that surgical science has progressed remarkably such that body/brain exchanges can be done without any pain or complications. So, here is the choice for you, GB: You can either keep your current Brain and have it transferred into a body that is a lot like yours except that it is a male twin/clone; OR you can keep your current Body, and have installed into it a Brain twinned/cloned from yours, except that this brain feels itself and knows itself to be female."

Procedure:
First, the experimenter will ask every participant to describe their feelings (in as much detail as possible) as the reality of the experiment is imagined, then the mood thermometer measure will be administered, followed by the gender content/discontent instrument. Then each participant will be asked: "Which is more your real-self, your me? Which remains as you—the original body or the original brain—and which is transplanted?"

Analysis

1. Look at group statistics—which is more real-self, me? Does body or brain remain?
2. Compare Male-to-Female data with Female-to-Male data
3. Assess affect/mood measure, and gender content/discontent instrument
4. Compare Male-to-Female participants with Female-to-Male participants on affect/mood and gender content/discontent measures.

PARTS ONE AND TWO

Analysis
Perform a number of paired statistical contrasts on Body vs. Brain as real-self:

1. Trans vs Controls

2. Male-to-Females vs. Female-to-Males
3. Cis-Male controls vs. Cis-Female controls
4. Male-to-Females vs. Cis-Male controls
5. Male-to-Females vs. Cis-Female controls
6. Female-to-Males vs. Cis-Female controls
7. Female-to-Males vs. Cis-male controls

Hypotheses

1. No clear hypotheses for (a) Trans group as a whole, (b) Control participants as a whole, or for (c) Male-to-Females, (d) Female-to-Males, (e) Cis-Males or (f) Cis-Females, each as a whole group.
2. No clear hypotheses on all the possible contrasts: (a) Trans vs Controls; (b) Male-to-Females vs. Female-to-Males; (c) Cis-Male controls vs Cis-Female controls; (d) Male-to-Females vs. Cis-Male controls; (e) Male-to-Females vs. Cis-Female controls; (f) Female-to-Males vs. Cis-Female controls; (g) Female-to-Males vs. Cis-Male controls
3. But the Trans participants who would keep the natal body and shift the brain might be more amenable to psychiatric treatment, while not being good candidates for re-assignment surgery. Also, perhaps other somatic treatments would also not be indicated for this group.
4. And there is a predicted affect and mood difference in which Trans participants—especially those with higher discontent scores on the gender content/discontent instrument—would feel less dread, and more positive affect then their matched Cis control participants. A more positive affect and mood might be indicative of better prognosis for future interventions.

Future Directions

Brain correlations with any or all findings would be fascinating. (This of course would be far down the road, involving much more expertise, time, money etc.)

The brain comparisons of interest would include:

1. Those participants who felt Body was more *Self* (Body-Ps) vs. those who felt Brain was more *Self* (Brain-Ps)
2. Trans Body-Ps vs. Cis Control Body-Ps;

3. Trans Brain-Ps vs. Cis Control Brain-Ps;
4. Cis Control Body-Ps vs. Cis Control Brain-Ps;
5. Trans Body-Ps vs. Trans Brain-Ps—This would perhaps be the most interesting and important comparison.

Note that looking at brain data for Brain-Ps vs. Body-Ps within each separate group would also be of much interest, perhaps providing less noisy data:

1. Male-to-Females: Body-Ps vs. Brain-Ps
2. Female-to-Males: Body-Ps vs. Brain-Ps
3. Control Cis-Males: Body-Ps vs. Brain-Ps
4. Control Cis-Females: Body-Ps vs. Brain-Ps

Experimental Philosophy Experiment Two: Body/Brain Exchange for Adults

Same Parts One and Two, as for Experiment One, including same Design, Analyses, Hypotheses, and Future Directions, except the Participants are adults.

Participants

PART ONE—Cis-Gendered

100 Cis Control Participants—50 Males; 50 Females—adults matched in age, socio-economic, and psychological factors to the Trans-Gendered Participants of Part Two 50 Males

Participants

PART TWO—Trans-Gendered

100 Trans adults—ages above 18, and all untreated medically with respect to gender.[5] 50 Male-to-Females; 50 Female-to-Males

[5]It might be difficult to find 100 untreated Trans adults. Perhaps then the solution would be to have two groups, one with treatment, one without, with different predictions for each group. Thus, those who have had bodily treatment might be more likely to see the real-self in the body exchange, than would those who have reached adulthood without such medical interventions.

2.2 The Real Continuer Experiment Plus Tele-Transporting—Experiment Three

Overview

This next Experimental Philosophy empirical study derives largely from Robert Nozick's (1981, pp. 33–70) classic Thought Experiment, involving what he called "Closest Continuer," assisted by Derek Parfit's (1986, pp. 199–209; 231–243) thought experiment vehicle—Tele-transporting. Whereas these classic works of Nozick and Parfit (discussed in Chapter 3) sought to address the question—Which *Self* (or aspects thereof) ought we be concerned with, when we consider future survival?—the aim in this monograph is different. Hence, in the current empirical Thought Experiment study, all the participants, children or adolescents, are asked questions designed to reveal their explicit and implicit notions of what counts as their "Real"-*Self* as they imagine themselves 15 years in the future, in other words their Closest Continuer when they are adults.

To begin the participants are told to imagine the following fanciful story as though it were real: "Our planet is about to be wiped out. Luckily you get to continue your life on a different planet, as the real you, but there is a catch—there will be a 15-year time lag. In other words, you'll continue as you but 15 years older. So, we'll now present you a series of paired pictures and your job will be to tell us each time which of the two pictures presented is closer to the Real You, but 15 years in the future."

All of the pictures will be modified to add 15 years to the participants' appearance. That is a control feature of the experiment. In the experimental condition there is another modification added—manipulating the photos so that they represent different degrees of gender change—from the extreme of Natal-100 percent/Trans-0 percent, through a middle ground of Natal-50 percent/Trans-50 percent, to the other extreme of Natal-0 percent/ Trans-100 percent. In this way, depending on which percentage gender change the participants deemed more "Real-self/Closest Continuer" the results could readily deliver diagnostic data relevant for future treatment planning. Also, findings could offer evidence for (or against) the new ontological phenotype, consisting of Trans-women and Trans-men. Were this to be a new category, it would exist within the standard binary gender system, and yet these phenotypes would be different from Men and Women, simpliciter.

To make this clearer with an example, let's suppose our reference subject, GB, were a participant in this experiment—this, before undergoing any transitioning. Would GB find a 15 years older Continuer with all GB's brain/mind content, but with a body quite similar to GB's but 100 percent male, paradoxically a Closer Continuer of Real-self, than a Continuer that was just like GB as she is—identical in all ways except fifteen years older? If so, this might be an indication for future physiologic transitioning. Suppose on the other hand GB responses indicate that GB views the 15 years older Natal-33 percent/Trans-67 percent modification as the Closest Continuer/Real-self. Then this would be some evidence favoring the new Tran phenotype—Trans-man in GB's case.

As with the earlier Experimental Philosophy studies, having presented the basics of the experiment, what follows is the presentation of it in empirical format, along with some plans for follow-up studies and future directions.

The Experiment—Presented in Empirical Format

Experimental Philosophy Experiment Three: Real Continuer plus Tele-Transporting

Design

Participants:
100 as yet untreated Trans children and adolescents. 50 Male-to-Females; 50 Female-to-Males

Set up:
Participants are told the following story: "Our planet is about to be wiped out. Luckily you will get Tele-transported to a different planet, getting to continue your life as you. But there is a catch—there is a 15 year time lag for this sort of interplanetary travel. So, this is why we'll now show you a series of paired pictures and your job will be to tell us each time which one of the two pictures presented is closer to the Real You, 15 years in the future."

Items:
The participants will be required to choose one of the pictures over the other as the more Real-self in each presentation—a so called forced choice design. The pairs will come from three sets, never two from the same set, and they will be presented in randomized order. The three sets are:

A. 40 different views of the individual participant in Natal-gender—all with 15 years added with a program that generate such images
B. 40 views of the individual participant with Natal-gender morphed 66.6–75% toward New, i.e. the Trans-gender—again all with 15 years added, by same program that can generate these types of images
C. 40 views of the individual participant with gender morphed 100% to Trans (New)-gender

The technical aspects of this empirical study will require finding a program that can generate both realistic gender-morphing and aging images based on photographic images of a human being. The Feusner et al. (2017) study described in Chapter 6 (text and footnote 6) has such a program—the "body-morph test"—which could be obtained.

Procedure:
Participants will be presented the 120 items paired in combinations of pictures from Set-A vs. Set-B; Set- A vs. Set-C; and Set-B vs. Set-C. This will be done twice for a total of 240 presentations. Each time, each participant, in the forced choice design, will be asked to choose the picture closer to the Real-self plus 15 years.

Analysis

1. Which of the three potential selves is the Real-self/Closest Continuer for each participant? This can be determined after statistically analyzing the 240 randomized forced choice presentations of pictures from Set-A vs. Set-B; Set-A vs. Set-C; Set-B vs. Set-C. Note that the number of presentations and the randomizing will allow for each participant's implicit as well as explicit ideas to be discerned.

2. A bar graph showing how many participants at each of the three different percentages of gender-morphing (all plus 15 years) deemed that one to be the Real-self/Closest Continuer This should be done overall for all participants and then for Male-to-Female participants and Female-to-Male participants separately, to pick up any differences. (There is no hypothesis regarding differences at this point.)

Hypotheses

1. Empirical support for a new ontological Trans phenotype would be gained if the statistical analysis demonstrated that at least some participants chose as their projected most Real-self (plus 15 years), photos from (Set-B) with gender 66.6–75% Trans morphed, over both (Set-A) with gender at 100% Natal, and (Set-C) with gender 100% Trans morphed. This new ontological phenotype category would be within the ordinary binary gender system—but persons whose projected Closest Continuers are from Set-B would perhaps best be considered Trans-men and Trans-women, rather than Men and Women simpliciter.[6]
2. If Set-A (100% Natal-gender) is chosen as the most Real-self/Closest Continuer (plus 15 years) by participants more than Set-B (66.7–75% Trans-gender morphed) or Set-C (100% Trans-gender morphed), these participants might be part of the cohort reported upon by Hahn et al. (2015, p. 3531; See footnote 2). There, the authors claim that around 75% of Trans children and adolescents do not pursue Trans-ness later, this despite the fact that all had been diagnosed with Gender Dysphoria at some point. Although there might be many reasons for this—some not reflective of the wishes of the

[6]Following up on this, those who chose Set-B could be recruited for a follow-up study—a somewhat different version of the Real-self/Closest Continuer experiment. Here, briefly is the set up and the question, with GB as a sample participant: "Life forms of all kinds are stopped. A re-start is scheduled within a week, and as part of it you are asked the following question: GB, now which of these three would be the Real-You for you: (a) A Male person who is like a twin, almost a clone, thus very similar although not identical with you in brain and body; (b) a GB-clone of you, identical to you as you are now in body and brain; (c) you, with whatever real world gender changes are possible, now instantiated, i.e., GB plus real world Trans-gender changes." Would choice C demonstrate more evidence for a separate ontological phenotype for transitioning/transitioned selves—in this case GB as a Trans-man?

child/adolescent—individualized data from an experiment like the Real Continuer plus Tele-transporting could help with future treatment plans.

Future Directions

Brain correlations with any or all findings from this empirical Thought Experiment would be fascinating. Again, this would entail much more expertise, time, money etc. However, given that there are some 25 or so recent (late 1990s to 2018) brain studies,[7] and because most show different sorts of brain differences in Trans persons—some involve white matter; some grey; some show changes in connectivity based tracts, i.e. areas concerned with "awareness of self," and yet others demonstrate cortical thickness variations—correlations with empirical responses from this Real-self/Closest Continuer X-Phi study should be most welcome. Minimally such correlations might help to focus future studies. There is a more ambitious goal too. We have seen, that when taken all together, the brain studies reveal an intriguing overarching question—cause or effect?—viz., is it the case that the brain differences discerned cause the feelings of Trans-ness; or have years of Trans experiences, felt and lived, changed the brain? Adding brain correlations to X-Phi studies like this one, in particular from children and adolescents, the youngest of whom have had fewer years experiencing Trans-ness, could offer highly valuable and useful information.

2.3 The Tele-Transporting Plus Percentage Replacement Experiment—Experiment Four

Overview

This Experimental Philosophy empirical study is also adapted from the work of Derek Parfit. (Cited just above and See Chapter 3). Interestingly, however, the aim of the current X-Phi Thought Experiment study is almost diametrically opposite from that of Parfit in the original version

[7] Many of these brain studies are discussed in Chapter 6.

of his classic Thought Experiment. Parfit wanted to explore the following question: "How much body and/or brain change (measured in percentages) could a person imagine enduring, and yet continue to regard the same *Self* as remaining?" Then, demonstrating that this question provided results that were hopelessly vague, he argued that the question itself was unremittingly problematic, thereby advancing his view that "Relation-R"[8]was more important than Personal identity when considering one's own future survival. The empirical X-Phi Thought Experiment presented here, on the other hand, primarily investigates how much gender-related change (also measured in percentages) would constitute a more satisfactory *Self* for Trans persons, while also supplying data on how much such change can be tolerated by the Cis-control participants.

The participants in this study are given a version of the Tele-transport narrative: "Earth is to be destroyed. Fortunately, you get to be re-assembled on a different planet. However, the technology is imperfect. Thus, the reconstituted versions of you will be somewhat different from you now, in various features and by different percentages. Now, in the experiment, you'll be given a series of pictures representing variations of reconstituted-you—i.e., pictures showing you-as-you-are-now, but not exactly, in that there will be varying degrees of change. Your task will be to tell us each time, with each presentation, whether the picture is the Real-You—yes or no."

The items will be pictures of each participant, gender-morphed to various percentages toward the New or Trans-gender. There will be included pictures that do and do not feature obvious gender markers, such as clothing and hair style; and pictures that do and do not include genitals and breasts.

There are three hypotheses, all concerning differences between the importance of obvious gender markers and genitals. First, we hypothesize that at the group level, both genitals and obvious gender markers will be more important in characterizing the Real-self for the Cis-control participants vs. the Trans participants. The next hypothesis suggests that

[8]Recall that Relation-R comprises psychological connectedness and psychological continuity, and that for Parfit, a physicalist, the psychological is brain-based.

the Trans participants actually consist of two groups, one of whom does, and the other of whom does not find the genitals (and breasts) significantly determinative of *Self.* The third hypothesis, based on the second one, suggests diagnostic and prognostic implications for both groups of Trans participants. Namely, those who find the genitals etc. determinative might be candidates for re-assignment surgeries, while the other group of Trans individuals might either (a) represent the percentage of younger Trans persons found not to pursue the Trans experience into adulthood; or (b) provide evidence for the new ontological phenotype category—Trans-man and Trans- woman. In both cases, surgical reassignment procedures may not be necessary for improved life satisfaction.

As with the prior experiments, the empirical details of Experimental Philosophy study number four, along with ideas for future directions, follow immediately below:

The Experiment—Presented in Empirical Format

Experimental Philosophy Experiment Four: Tele-Transporting plus Percentage Replacement

Design

Participants:

100 as yet untreated Trans children and adolescents 50 Male-to-Females; 50 Female-to-Males

100 Matched Control Participants (matched for age, socio-economic, and psychological features) 50 Cis-Males; 50 Cis-Females

100 Trans adults—ages above 18, and all untreated medically with respect to gender.[9] 50 Male-to-Females; 50 Female-to-Males

100 Matched Control Participants (matched for age, socio-economic, and psychological features) 50 Cis-Males; 50 Cis-Females

[9]As discussed for the prior experiment, it might be hard to find 100 untreated Trans adults. Please see footnote 5 (this chapter) for a partial solution.

Set up:
Participants (Ps) are told the following story: "Our planet is to be wiped out. Luckily you get to be re-assembled on another planet—except that the technology is imperfect. Thus, you will be given a series of pictures of re-constituted versions of you, and your task will be to tell us which are close enough to the Real-You to count as you for you."

Items:
There will be pictures of Participants generated with 5%, 10%, 20%, 40%, 60%, 80%, or 100% morphing to the New/Trans-gender

2 sets of such pictures, each with 10 pictures, will be constructed:

A group—Pictures of P in Natal-gender at 90%, 80%, 60%, 40%, 20%, 10%
B group—Pictures of P in Trans-gender at 90%, 80%, 60%, 40%, 20%, 10%
Each group will have 4 extras—what I'll call "genital" pics (but to include breasts where apt).
Natal at 95% with Natal genitals; Natal at 5% with Natal genitals
Natal at 95% with Trans genitals; Natal at 5% with Trans genitals
Trans at 95% with Natal genitals; Trans at 5% with Natal genitals
Trans at 95% with Trans genitals; Trans at 5% with Trans genitals

There will be overlap here, but that is a positive feature of the design, in order to get at implicit as well as explicit constitutions of *Self*.

Each P will be shown 80 different pictures as the 2 sets (A and B) are presented in 4 different conditions:

I. Pictures as gender neutral as possible—no obvious cues with clothes, hair, etc.—No genitals
II. Pictures as gender neutral as possible—no cues—Except yes genitals
III. Pictures rife with gender cues—No genital
IV. Pictures rife with gender cues—Yes genitals

Note that this experiment too will require expertise in photo morphing for the item design. The Feusner et al. (2017) study's body morph program (referenced earlier in this Chapter and in Chapter 6) would seem adequate.

Procedure:
For each participant, the 80 different pictures will be presented 4 times in randomized order for a total of 320 presentations. For each of these presentations, the participants, in a forced choice design, will respond "Yes" or "No" to the question—Real-You? Yes or No?

Analysis

1. For each participant, calculate at what percentage the Real-self is constituted, in each condition.
2. Bar graphs of number of participants at which percentages in every condition can be constructed, with comparisons of every participant-type as in #3 just below.
3. For importance of genitals compare sets I + III vs II + IV.
   For importance of other cues compare sets I + II vs. III +IV.
4. For every comparison look at:

   a. Trans vs Controls
   b. Male vs. Female Controls
   c. Male-to-Females vs. Female-to-Males
   d. Male-to-Females vs. Male Controls
   e. Male-to-Females vs. Female controls
   f. Female-to-Males vs. Female Controls
   g. Female-to-Males vs. Male Controls

Brain correlations wherever possible would be great, with any group or response effects linked to brain findings a real plus.

Hypotheses

1. For the Cis-Control participants, the genitals will have more determinative power, i.e. Control Participants will say, "No, not real-self," even with Natal at 95% as long as there are Trans genitals.
2. This will be markedly different for at least one subset of Trans participants in which genitals are not significantly determinative. If

such a subset exists, this would be a group for which re-assignment surgery would not be necessary, and in fact contraindicated. Moreover, this group may provide empirical evidence for new ontological phenotype categories—Trans-men and Trans-women—still within the mainstream binary gender system, but additional, and distinct from Men and Women as phenotypes. Or, members of this cohort may be among the purportedly high percentage of Trans persons who do not pursue trans-ness into middle adulthood. (See again footnote 2.)

3. There might be another Trans participant subset where genitals are the drivers, i.e., "Yes, Real-self" with as much as Natal 95% as long as there are Trans-genitals. If this group is indeed found to exist, these individuals might prove good candidates for later re-assignment surgery.

Future Directions

As mentioned above, brain correlates would be most welcome. Also, follow-up with respect to the predicted hypotheses about the distinct groups of Trans persons would be of interest and importance.

3 Moore's Paradox: Addendum to the X-Phi Empirical Experiments

The notion that Trans persons might find certain sorts of Moore-Paradoxical statements assertible with neither incoherence nor irrationality was initially the trigger, the catalyst for this work. And so, it is fitting to return to Moore's Paradox, placing it among the Experimental Philosophy studies. As was discussed in the opening chapter of this work, it is debatable whether or not Moore-Paradoxical statements obtain for Trans persons. I will not repeat the different views on this matter here; however, there is a way in which Moore-Paradoxical statements could be of use in all of the Experimental Philosophy empirical studies outlined above. With no extra equipment, little added time, and minimal additional statistical testing, each Trans participant in every experiment could be queried as to whether or not they would endorse the relevant

Moore-Paradoxical statements. Various correlations and group analyses could readily be performed—(a) Male-to-Females vs. Female-to-Males; (b) age groups; (c) those Trans persons who located the incongruence in the brain vs. those whose body seemed "wrong;" (d) Trans participants who fit well into the new ontological phenotype categories of Trans-woman or Trans-man vs. those remaining in the standard Man or Woman; and (e) certainly any correlations with any of many potential brain findings.

To outline the procedure and questions, let's once more take GB, the Trans person, who has not (yet) transitioned, we've employed as representative throughout. Here is what GB would unfold in having GB address these questions[10]:

Starting, and including the uncontroversial statement: "I have a female body"—to which you have agreed; would you please indicate whether each of the following statements is true or false?

"I have a female body."——————————-True or False
"[So] I am female, but I don't believe that." ——True or False
"[So] I am not male, but I believe I am."——— True or False

The first statement should be endorsed by all Female-to-Male participants, who have not (yet) transitioned.
The second one is a Moore-Paradoxical positive statement.
The third one is a Moore-Paradoxical negative statement.

One might predict that the second and third statements would be found true to the same extent. However, with the possibility of the new ontological phenotype, there could be a substantial difference, one that is important to contemplate. Thus, GB could, with all sincerity, not be able to assert positive statement two, while endorsing the Moore-Paradoxical negative statement three: "I am not male, but I believe I am male—a Trans-male type male."

All of these are empirical questions, and worth investigating.[11]

---

[10]The Male-to-Female version would be the same, except with the gender words reversed.

[11]There are at least two other GB responses that would not be Moore-Paradoxical. The first one, given in the Introduction Chapter is: "Even though it is true that I have a female body"…

4 An Admission Regarding the Experimental Philosophy Studies

These empirical adaptations of classical philosophical Thought Experiments were designed to provide useful data and findings. The Experimental Philosophy studies presented, could, if successful, support several of the hypotheses offered, including: (1) New ontological phenotype categories—Trans-men and Trans-women, these in addition to Men and Women, all within the standard binary gender system; (2) Diagnostically important information about Trans participants based on different outcomes in the various studies; (3) Relevant data for the direction of future treatments (if any treatment is indicated), and prognoses therein; and (4) Potential correlations with brain findings, with the promise of possibly gaining some ground on cause vs. effect questions.

However, even if these investigations were to prove amazingly productive, they will have not answered our initial three fundamental philosophical questions about the *Self*—the ontological, epistemological, and conceptual questions arising owing to the Trans phenomenon. Thus, in all truthfulness, I now report that the proposed Experimental Philosophy (X-Phi) empirical studies must join the accounts of the groups explored throughout this project—(1) The classical philosophers; (2) The modern philosophers; (3) The gender theorists; and (4) The brain scientists—in having gained insufficient traction on these questions toward proposals for new conceptions of the *Self*. Yes, even if one considers the work of all of these diverse groups, together, and/or in various combinations, we still don't have a satisfactory answer to the ontological question: namely, what *Self* or part thereof is involved in the Trans incongruence? Nor, can we answer the epistemological puzzle: How does the *Self* know of

"I know I am a boy; but I can't (quite) believe it." Again, this is not a real Moore-Paradoxical statement because "I can't (quite) believe"—i.e., not being able to believe, is different from the starker attitude of not believing— "I don't believe." Also note the qualifier, "quite." The other response that is not Moore-Paradoxical starts similarly "Even though it is true that I have a female body"..."I know I am not female, I am male." This is not a Moore-Paradoxical insofar as for persons making this particular statement, having a female body has not been equated with femaleness at all.

the gender inconsistency, and with what evidence? Finally, the conceptual notion of the *Self* that could accommodate these answers is still out of reach.

But, despite this sad acknowledgement, in the attempt to triangulate from these many angles in order to solve the mystery, it is happily the case that this exploration has covered much interesting territory. And so, with a mixture of emotions, including both disappointment and satisfaction, let me move finally to the Summary and Conclusions chapter, there to re-visit briefly what we can, and cannot, rightfully conclude.

Bibliography

Feusner, J., Lidstrom, A., Moody, T., Dhejne, C., Bookheimer, S., & Savic, I. (2017). Intrinsic network connectivity and own body perception in gender dysphoria. *Brain Imaging and Behavior, 11,* 964–976.

Hahn, A., Kranz, G., Kublbock, M., Kaufmann, U., Ganger, S., Hummer, A., et al. (2015). Structural connectivity networks of transgender people. *Cerebral Cortex, 25,* 3527–3534.

Nagel, T. (1986). *The View from nowhere.* Oxford: Oxford University Press.

Nozick, R. (1981). *Philosophical explanations.* Cambridge, MA: Harvard University Press.

Shoemaker, S. (1963). *Self-Knowledge and Self-Identity.* Ithaca, NY: Cornell University Press.

Williams, B. (1970/1973). *The self and the future. In problems of the self* (Chapter 4, pp. 46–63). Cambridge: Cambridge University Press.

Zavala-Rojas, D. (2014) Thermometer scale (feeling thermometer). In A. Michalos (ed.), *Encyclopedia of quality of life and well-being research.* Dordrecht: Springer. https://doi.org/10.1007/978-94-007-0753-5.

8

Summary and Conclusions

Abstract Disappointments and accomplishments are outlined. Moore's Paradox dilemma was not solved, nor the three fundamental philosophical questions occasioned by the Trans phenomena—the ontological, epistemological and conceptual *Self* questions. This, despite wide ranging explorations. However, the many sorts of investigations, and the interspersed Contrast Cases, have allowed a proposed list of features—necessary for, but not necessarily sufficient to constitute a human *Self*. Moreover, a proper-function account of the Trans-gender phenomenon, linking philosophical concerns with biological matters has been advanced. And, a new ontological phenotype has been proposed. Finally, the empirical Experimental Philosophy studies, adapted from classic thought experiments about the *Self*, could be diagnostically useful for Trans persons, as well as potentially providing evidence for or against the new phenotype hypothesis.

Keywords *Self*: Trans vs. Cis · Moore's Paradox · Three foundational questions: ontological, epistemological, conceptual · Evolutionary proper-function account of gender/Trans-gender · X-Phi empirical studies · New ontological phenotype: Trans-men/Trans-women

L. A. W. Brakel, *Investigations into the Trans Self and Moore's Paradox*,
https://doi.org/10.1007/978-3-030-44645-1_8

This project began when I was asked whether or not Trans individuals might be able to make Moore-Paradoxical statements without incoherence or irrationality.[1] Thus, could a girl who wants to, and will transition state: "I am a girl, but I don't believe that"—the positive Moore-Paradoxical statement; or "I am not a boy, but I don't believe that"—the negative version? The answer here is not clear. Some Trans persons, who are natally gendered as girls, feel they've always "known" they are male, not female—so that neither version of Moore-Paradoxical statements would work. On the other hand, Trans theorists have observed that this sort of early certainty is often at least partly political, necessary in part to receive a medical diagnosis, which itself necessary as prelude to medical/surgical transitioning procedures.[2]

But this question was just the first of three additional and more fundamental philosophical questions—all centering on how the Trans experience might/could/should help provide a more complete understanding of the philosophical concept of *Self*. So, first there is the ontological question: What *Self* or aspect of a person's *Self* knows that he/she is the "wrong" gender? Next comes the epistemological question: How does a person know that he/she is the "wrong" gender; and what does he/she know? Finally, the conceptual question: In light of the first two questions, in what way should we change and/or perhaps expand our concept of *Self*.

Almost immediately the project took off in a non-linear fashion—more precisely, it launched as would several intersecting rays of thought, emanating from various home disciplines, in different directions, and all at once. One such ray, consisted of assembling various "Contrast Cases." The Contrast Cases were selected in that they were interestingly both similar and different from the Trans case. The similarities were of various

[1] My husband, Arthur Brakel, is the person responsible for the initial monograph-sparking query.

[2] See Chapter 1, Footnote 5. Another type of political-social issue has concerned me in throughout this project. For heuristic purposed I chose to focus only upon those Trans persons within the mainstream binary gender system who felt that their natal gender was the "wrong" gender. These restrictions were imposed solely for the academic goal of investigating the most distinct and extreme cases. In short, I have nothing but equal respect for those who are mixed-gendered, fluid-gendered, non- or a-gendered, and third-gendered.

sorts, but one difference was always the same: The *Self* concept involved in each Contrast Case could be more readily accommodated by existing philosophical views of the *Self*, or combinations thereof, than could the Trans case.

Another ray pointed the way toward the exploration of three sorts of philosophical views of the *Self*: (a) four classic accounts—Descartes, Hume, Kant, and Freud; (b) several modern views, including those dealing with *Self*-constitution, and those concerned with the *Self* we care about when we care about our survival; and (c) selected views from gender-theorists. The investigations of these many and varied kinds of philosophical accounts of *Self*, in addition to the several Contrast Cases, suggested the following incomplete list of elements necessary (but not necessarily sufficient) for a human *Self*:

1. Bodily percepts, inner and outer registrations
2. Intuitions/Concepts
3. All of these both *in* a Human, and *being for* a Human being (under a concept/judgement)
4. TUA, the diachronic and synchronic synthesizer, expressing the uses of "I think"[3]
5. Agential actions
6. Memories (some somatic, some psychological)
7. Personal identity
8. Self-Constitution
9. Primary Consciousness
10. Reflective Consciousness
11. Self as object distinguished from other objects
12. Self as subject distinguished from other subjects (for Humans).

And yet nothing from this list suggests anything different about Trans individuals from Cis individuals that could help us with the three foundational, but continually puzzling, philosophical questions. Even with

[3]Although, numbers 3 and 4 refer to Kantian terms, the ideas have wider application. (See Chapter 2 for more on this.)

respect to bodily registrations, both internal feelings and outer sensations, seem unremarkable and fluid, with Trans persons demonstrating nothing significantly unusual or different.[4]

Meanwhile, there were some intriguing parallels hidden within the obvious contrasts among the various accounts of *Self.* I attempted to make good use of these parallels, particularly in the three more original chapters of this project. (I refer to Chapters 5–7.) For instance, noting that from unrelated quarters, notions of evolutionary fitness success and human flourishing contributed significantly to constituting *Self*-hood, I was able to develop an evolutionary proper-function selective fitness account for gender, including Trans-gender. Sex-changing fish were instrumental in providing a proof-of-concept anchor for the account. And a gender-theorist's convincing claim regarding idea of Trans-men and Trans-women, as non-marginalized, but instead representative of what could be construed as a different ontological phenotypic category within the binary gender system, not only helped shape the proper-function account constructed, but led to other areas for exploration.

Neuroscientific brain study research was one such area. Through investigating approximately 20 studies on Trans and Cis humans, done over the last 25 years, the question of a new ontological phenotype category could be addressed with empirical data from a physiological perspective. Here too proof-of-concept evidence from studies on the brain and behavior of sex-changing fish studies proved revealing and surprisingly salient. And the fish research also contributed to another brain-based Trans question—are Trans phenomena caused by brain differences; or are brain differences resultant epigenetic changes from lives lived as Trans persons; or is it a combination; or does it vary individual by individual?

In the penultimate chapter of this monograph, re-visiting a set of classic Thought Experiments—experiments that had been discussed earlier

[4]Here is one possible exception: A distinction between Trans and Cis persons might be demonstrable in both aspects of Parfit's Relation-R-*Self* (see Chapter 3). Recall that Relation-R, according to Parfit comprises both self-connectedness and self-continuity. It is plausible then that Trans persons would find more self-connectedness with their New/Trans-gendered versions of *Self* than any *Self* version were they not to transition. This would be all to the good. Similarly, with transition, a Trans person might experience less self-continuity with their troubled natal-gendered-*Self*, than if no transition took place. And less self-continuity could be a desirable outcome!

(Chapter 3) in the subsection on "the *Self* I care about when I care about my survival"—I devised empirical Experimental Philosophy (X-Phi) adaptations. These were constructed in order to address important questions relevant to Trans persons. For example, if it is a given that a Trans person feels an incongruity in their gender assignment, where is that located—in the Body or in the Brain/Mind? Experiments capable of providing empirical data leading to convincing answers to such questions could be invaluable as a source of diagnostic and prognostic information. Moreover, these studies were designed such that they also could provide significant evidence for or against the idea of a new ontological phenotype category—Trans-men and Trans-women—within the binary mainstream system.

Yet, on the whole, I must confess that in this short monograph I have raised many more question than I have been able to answer. I have neither found nor proposed a conceptual change in the philosophical notion of the *Self*—one that could answer (or even make acceptable accommodations to) the foundational questions about the nature of *Self* raised by the Trans experience. Indeed, as these epistemological, and ontological *Self* questions remain vexing, so too does the seemingly needed conceptual change remain out of reach, at least for now (and certainly for this author). Also, the Moore's-Paradox puzzle remains largely unsettled. I say largely, because, were the new ontological phenotype notion to go through, someone like GB could state the following negative type Moore-Paradoxical sentence with neither irrationality or incoherence: "I know I'm not a boy [i.e., boy, simpliciter], but I don't believe that…because I am a boy, a Trans-boy."

Still, that is but a thin reed. And so, for now, despite my efforts here, we must conclude that existing philosophical accounts of the *Self*, even in combination, are inadequate to account for the vexing issues encountered as we consider the Trans experience. But we should not despair. The recognition of incomplete philosophical analyses, like the failure of replication studies in empirical science, must be acknowledged in order to provide stimuli to future analyses with new understandings. Moreover, in terms of this monograph itself, in exploring the work of classic philosophers, modern philosophers, gender-theorists, fish researchers, and brain scientists, I have been able to advance an evolutionary proper-function analysis, derive and develop Experimental Philosophy empirical studies, and posit a new ontological phenotype. Perhaps these diverse achievements can function as useful bridges, bringing together insights from the many highly disparate areas traversed in this project.

Index

L. A. W. Brakel, *Investigations into the Trans Self and Moore's Paradox*, https://doi.org/10.1007/978-3-030-44645-1

The manufacturer's authorised representative in the EU is Springer Nature Customer Service Centre GmbH, Europaplatz 3, 69115 Heidelberg, Germany. If you have any concerns regarding our products, please contact ProductSafety@springernature.com

Printed and bound by CPI Group (UK) Ltd, Croydon, CR0 4YY

30/09/2026

02225475-0020